Solid and Hazardous Waste Management

NIPA GENX ELECTRONIC RESOURCES & SOLUTIONS P. LTD.
New Delhi-110 034

About the Editor

Dr. Sunil Kumar is a well-rounded researcher with more than 22 years of experience in leading, supervising and undertaking research in the broader field of Environmental Engineering and Science with focus on Solid and Hazardous Waste Management. Dr. Kumar is a graduate in Environmental Engineering and Management from Indian Institute of Technology, Kharagpur, India. He completed his Ph. D. under the joint supervisions of Dr. Sukumar Devotta (Former Director, CSIR-NEERI, Nagpur who got commendation letters from the Executive Director of UNEP and the President of Switzerland, and the Nobel Peace Prize in 2007 as a part of the United Nations Inter-Governmental Panel on Climate Change) and Prof. Somnath Mukherjee of Jadavpur University, Kolkata, India. His primary area of expertise is solid waste management (Municipal Solid Waste, Electronic waste etc.) over a wide range of environmental topics including contaminated sites, EIA and wastewater treatment. His contributions in these fields led to a citation of 10136, h-index of 48, and i10-index of 188 (Google scholar). His contributions since inception at CSIR-National Environmental Engineering Research Institute (NEERI), India in 2000 include more than 250 refereed publications, 05 books and 40 book chapters, 10 edited volumes and numerous project reports to various governmental bodies and private, local and international academic/research bodies. He is the Associate Editor of peer reviewed journals of the international repute i.e., Environmental Chemistry Letter, International Journal of Environmental Science & Technology, ASCE Journal of Hazardous, Toxic and Radioactive Waste. He also serves as Editorial Board in Bioresource Technology, Elsevier. He has completed more than 25 research projects as PI with 15 (7 awarded) Ph. D. and 20 M. Phil/M. Tech thesis/ dissertations. He was an Expert committee member for revision of Solid Waste Management Manual in India, NGT members for implementation of Solid Waste Rules in UP, India, organized national and international workshops/conferences and delivered many invited talks at both Indian and international Venues. Dr. Kumar was awarded as Outstanding Scientist in 2011 and 2016 at CSIR-NEERI for his Scientific Excellence in the field of Research & Development in Solid Waste Management. Dr. Kumar was also awarded with the most prestigious award Alexander von Humboldt-Stiftung Jean-Paul-Str.12 D-53173 Bonn, Germany as a Senior Researcher for developing a Global Network and Excellence for more advanced research and technology innovation.

Solid and Hazardous Waste Management

Sunil Kumar

Humboldt Senior Fellow
Sr. Principal Scientist and Head
Waste Re-processing Division
Professor, Academy of Scientific and Innovative Research
CSIR-National Environmental Engineering Research Institute (CSIR-NEERI)
Nehru Marg, Nagpur – 440 020, Maharashtra, India

NIPA GENX ELECTRONIC RESOURCES & SOLUTIONS P. LTD.
New Delhi-110 034

NIPA GENEX ELECTRONIC
RESOURCES & SOLUTIONS P. LTD.

101,103, Vikas Surya Plaza, CU Block
L.S.C.Market, Pitam Pura, New Delhi-110 034
Ph : +91 11 27341616, 27341717, 27341718
E-mail:newindiapublishingagency@gmail.com
www: www.nipabooks.com
For customer assistance, please contact
Phone: + 91-11-27 34 17 17 Fax: + 91-11- 27 34 16 16
E-Mail: feedbacks@nipabooks.com

ISBN: 978-93-90591-85-5

Composed and Designed by NIPA.

Foreword

Solid Waste Management (SWM) is one of the most crucial aspects of urbanization planning and infrastructure development. The governing authorities of India realized its gravity and launched a countrywide campaign, 'Swachh Bharat Mission (SBM)' in October 2014. Now, the time has come to look forward and accentuate various aspects of SWM through genuine assessment and delivering appropriate solutions to the problems associated with SWM. Thorough knowledge coupled with a deep sense of understanding regarding various aspects of solid waste engineering and management is the key factor to succeed in this field. The book entitled 'Solid and Hazardous Waste Management' covers all the features of SWM specifically focused on developing countries like India. The significance of comprehension about Municipal solid waste management (MSWM) has been put forward through several subtopics.

It elucidates various features associated with SWM, including Global waste stream scenario, Waste recycling strategies, Emissions from different waste streams, Environmental impacts of waste management techniques, Novel tactics for sustainably managing wastes, Legislations, and management practices of municipal solid wastes, etc. The active discussion and peer-reviewed write up on different waste management aspects have made the book perfect for academicians, researchers, students, and policymakers. I am confident that this book is beneficial to all stakeholders persistently working in solid waste management in developing countries and that it will provide discernment to others.

Surampalli Y. Rao, Ph.D., P.E.

Preface

Developing countries like India face various environmental issues associated with sustainable management of solid wastes. The current waste management system in Indian scenario seems inept considering different aspects including rate of waste generation, inadequate waste collection and transportation facility, processing, and disposal mechanisms. In the recent past, the Indian Prime Minister received the prestigious Global Goalkeeper Awards from the Bill and Melinda Gates foundation for launching the benevolent campaign 'Swachh Bharat Mission'. One cannot deny efforts taken by the Indian government for directing India towards Open defaecation free nation; however, the poor waste management infrastructure is the country's major shortcoming.

Even today, the major portion of the generated solid waste is directly dumped indiscriminately in the open spaces or in the outskirts of the urban or semi-urban regions without any prior treatment, thereby leading to contamination of groundwater, rise in air pollution problems and similar other environmental issues. The key aspect of an effective waste management system is ensuring appropriate segregation of waste at source and ensuring that the segregated waste goes through different streams of recycling and resource recovery. It is intrinsically necessary to offer formal guidance to the urban local bodies for making them effective and self-reliant in sustainable solid waste management in their respective regions.

The text in this book is organized into the areas important for all the environmental researchers working in solid waste management domain. Chapters on various aspects of solid wastes and its management comprising the Global waste stream scenario, Recycling of wastes, Strategies on management and mitigation of greenhouse gases, Various challenges and shortcomings associated with solid waste management in developing countries are included in this book. We have included the fundamental topics and tenets of solid waste management, explaining them various contemporary instances. We have also incorporated emerging aspects including the Techno-economic aspects of solid waste management in developing countries like India as well application of advanced technologies and novel tactics for sustainable waste management. Environmental education and communication as a foundation in sustainable development, an essential aspect for strict enforcement of various environmental policies, is also discussed in the present book.

We as authors and assistant editors are grateful to Dr. Sunil Kumar, Sr. Principal Scientist and Head at Waste Reprocessing Division in CSIR-National Environmental Engineering Research Institute (CSIR-NEERI), Nagpur who has worked immensely in the field of Solid and Hazardous Waste Management having more than 20 years of experience for giving us the opportunity to work on this book. Without his constant support and encouragement this book would never have come to fruition.

Sonam Paliya
Ashootosh Mandpe

Contents

1

Global Waste Stream Scenario

Suraj Negi[1], Aishwarya Rani[2], Chang-Tang Chang[2*] and Allen H Hu[1]

[1]Institute of Environmental Engineering and Management, National Taipei University of Technology, Da'an District, Taipei City 106, Taiwan, R.O.C.
[2]Department of Environmental Engineering, National I-Lan University Yilan City, 260, Taiwan

ABSTRACT

Waste is generated by almost every activity, and it holds no value to the first user(s). Solid waste has been creating nuisance much before air and water pollution came into the picture. The evolution of time, advancement in technologies, new services and products, has changed waste in terms of quality and quantity. The discarded materials on the curb every day; creating a massive heap of solid waste. The waste stream is a significant challenge for developing as well as developed countries. For managing the huge volume of waste, recycling and composting are being practised. However, the most preferred way is landfill and/or incineration. The global view of the waste stream is a troubling and alarming picture. This chapter focuses on the waste stream scenario at a global level and gives a clear representation of the waste collection system. This chapter also covers the waste stream scenario on a global scale, which is an essential and foremost step for effective waste management.

Keywords: Solid waste, Global view, Waste Collection, Waste Composition, Waste Categorization

1.1. Background Introduction

We live in a world with "throw-away" societies that are involved in a pattern of consuming packaged products that don't even last past a single use or even a year and discarding the waste. This lifestyle generating waste is negatively impacting public health, environment and imposing socio-economic problems on us. The waste is continuously consuming our land, resources, and energy for storage, management, treatment, and disposal. Solid waste was an urban issue initially but has reached every corner nowadays. The composition and generation rates of solid waste vary from country to country. It also varies with the sources such as households, offices, restaurants, markets, shops, public institutions, waterworks, sewage facilities, agricultural activities, construction and demolition sites. Globally , the waste is booming in weight and becoming complex in nature with the escalating rate of urbanisation accompanied by the hike in population for making its management and treatment intrinsic. Mixed solid waste is the by-product of urbanisation. In the later sections, the waste categorisation is discussed in detail.

1.2. Waste Categorisation

Solid wastes are organic and inorganic waste materials, which involves furniture, kitchen refuse, product packaging, grass clippings, batteries, paint cans, paper and appliances. Either a heterogeneous mixture of waste from an urban community or a more homogeneous accumulation of industrial, mineral and agricultural wastes, solid waste can be any or both of it (Kumar et al., 2018). The waste generation on the global scale would reach from 3.5 million tonnes per day in 2010 to 6 million tonnes per day in 2025, i.e., from 1.2 kg/capita/day to 1.42 kg/capita/day (Hoornweg and Bhada, 2012). Islands are generally observed to have more waste generation per capita per day due to waste generated by the tourism industry. East Asia and the Pacific region generate 270 million tonnes of solid waste; 70% of which is contributed only by China (Hoornweg and Bhada, 2012).

Waste categorization is the critical concept for waste segregation reducing, reusing and recycling. When the government took the solid waste collection and management responsibilities, it was only limited to residential and commercial activities. Complete understanding and knowledge of the categories of solid waste and their sources of generation is important, in addition to the composition and the rate at which it is being generated and disposed of (Manyapu et al., 2019, Kumar et al., 2017, 2009; Wainaina et al., 2020which are facing a serious pollution problem due to the generation of huge quantities of solid waste. This paper presents an assessment of the existing situation of municipal solid waste management (MSWM). It plays an indispensable role in effective solid waste management as well as in the design and operation of the associated functional elements. Based on the nature of waste, solid wastes can be broadly categorized into municipal solid wastes, biomedical wastes, agricultural wastes, hazardous wastes, electrical and electronic wastes, and other categories, which includes construction and demolition wastes, scrap tires, batteries, used oil, sanitary waste and radioactive and nuclear wastes. These all are discussed in the following subsections (Section 1.2.1 to 1.2.6.).

1.2.1. Municipal Solid Wastes

Municipal solid wastes consist of waste generating from household activities. Globally, 1.3 billion tonnes of solid waste is generated annually (Hoornweg and Bhada, 2012). Out of the generated 122-136 million tonnes of solid waste in India, treatment is provided to only 20% (CPCB, 2017). It comprises of food wastes, plastics, cardboard, yard wastes, leather, textiles, glass, wood, metal, ashes, *etc*. Some of the solid waste management practices are source reduction, collection, recycling, composting, incineration, landfilling or dumping. The treatment of municipal solid waste becomes more challenging due to the lack of segregation at the source. The highest fraction of the municipal budget is spent on solid waste management. The ineffectiveness of municipal solid waste management is caused due to the less or no practice of segregation at source.

Some civic authorities have started segregating waste at the source due to the high waste generation and less disposal area. In India, municipal solid waste dominates the biodegradable portion (Mandpe et al., 2020a).

1.2.1.1. Packaging Waste

The first plastic was created in the early 20th century. Since then, the hydrocarbon molecules are being cut up and repurposed to make up the plastics family. From logistics to food and from clothes to cosmetics, everything is covered by a layer of plastic or paper packaging. Day by day, it shares a large portion of our municipal solid waste. Out of manufactured plastics in India, 43% of plastics is used for packaging dominated by single-use plastics (TERI, 2018). About 267 species are affected by plastic debris globally, which includes 44% of seabirds, 86% of sea turtles and 43% of marine mammals (TERI, 2018). In 2016, 221 kg and 55 kg of packaging waste per person were generated by Germany and Croatia, respectively (Packaging waste statistics, 2019). In India, 60 major cities generate around 25,940 tonnes of plastic waste per day (CPCB, 2017). It mainly comprises of recyclable polyvinyl chloride (PVC) and polyethylene terephthalate (PET), and non-recyclables, including multi-layer thermocol, fibre reinforced plastic (FRP) and sheet moulding compound (SMC).

As per the report by Implementation of Plastic Waste Management Rules (2016), Maharashtra and Gujarat are responsible for half of the 1.6 million tonnes of plastic waste generated per year in India. As per an estimate by the Federation of Indian Chambers of Commerce and Industry (FICCI), there are 7,500 recycling units in India. Half of them are unorganised. Single-use plastics are discarded in drains and rivers followed by oceans or dumped into landfills. In 2011, India took the first initiative to tackle the menace of plastic waste by notifying the Plastic Waste (Management and Handling) Rules 2011. Later, the more robust Plastic Waste Management Rules, 2016, replaced the former one. According to FICCI, Indians consume 11 kg of plastic per year, while Americans consume 109 kg of plastic per year.

1.2.1.2. Food Waste

The amount of food consumed by the United Kingdom is the amount of food wasted in India. As per the United Nations Development Programme, up to 40% of the food produced in India is wasted, and 21 million tonnes of wheat is wasted. 25% of freshwater and 300 million barrels of oil are used to produce food and are ultimately wasted (CSR Journal, 2018). Food waste generates about 8% of global greenhouse gas emissions (The Hindu BusinessLine, October 30, 2016). According to the Food and Agriculture Organisation (FAO), one-third of food produced, which is 1.7 billion tonnes, is wasted every year worldwide. Around US$ 1 trillion, US$ 700 billion and US$ 900 billion are the associated costs of economic, environmental and social loss (The Hindu BusinessLine, October 30, 2016). 24% of the food is wasted during production, handling and storage, followed by 35% consumption. This issue is of more concern as India ranks at 97 among 118 countries in the Global Hunger Index of 2016.

It takes a landmass larger than China to grow food each year that is ultimately never eaten. When food waste goes to landfills, it decomposes anaerobically and produces methane which is 23 times more deadly than carbon dioxide. In the developed world, more than 50% of the food is wasted in homes and less than 2% in supermarkets. However, supermarket practices are directly responsible for food waste elsewhere in the supply chain. If the food waste is composted instead of landfill, the carbon dioxide would be emitted, which is a less harmful greenhouse gas. The ugly production, confusion in expiration dates, and habit of buying more are some causes of food waste.

1.2.1.3. Yard Waste

Yard waste is also known as vegetative waste which includes tree trimmings, leaves, grass clippings and other plant materials. It varies with the season and depends on the neighbourhood. The leaves, grass clippings and wood may account for 5-30%, 10-20% and 5-10% of the municipal solid waste, respectively (Environmental Encyclopaedia, 2003). Grass and leaves are the major components of yard waste by weight and by volume, respectively. Yard waste can't be ignored as it is the third-largest component of generated municipal solid waste by weight (Miller, 2013). In the United States of America, about 32.9 million tons of yard waste is generated per year, out of which 21.3 million tonnes are composted, and the rest is landfilled.

1.2.1.4. Glass

Glass is one of the components of municipal solid waste. It is found in the form of beer, wine and liquor bottles, soft drink bottles, jars and bottles for food, cosmetics and other products. In 2015, 10.43 million glass waste was generated in the United States, and it was 4.4% of generated municipal solid waste (Environmental Protection Agency (EPA), 2015). Glass waste is either recycled, burnt or landfilled. According to EPA (2015), 1.36 million tonnes of glass was combusted, and 6.35 million tonnes was landfilled in the United States.

According to the Federal Ministry for the Environment, Nature Conservation and Nuclear Safety, Germany, use of waste glass reduce the requirements of energy just by adding about 0.2-0.3% point of waste glass. Recycling waste glass is beneficial for the environment as it cuts carbon dioxide emissions and is diverted from the landfill space; thereby giving room to other wastes.

1.2.2. Biomedical Wastes

According to World Health Organization, medical waste is defined as "All waste materials generated at health care facilities, such as clinics, hospitals, blood banks, dental practices, physicians' offices, and veterinary hospitals/clinics, as well as laboratories and medical research facilities are termed as biomedical waste." All over the world, the waste generated in hospitals and pharmaceuticals consists of 80% of general waste and 20% of hazardous waste that may be of toxic, radioactive or infectious nature (Janagi et al., 2015). Biomedical waste comprises of infectious waste (swabs, bandages, cultures, gloves, blood and body fluids), hazardous wastes (chemicals, instruments, sharps), pharmaceutical waste and radioactive wastes from cancer therapies. Due to insufficient and inadequate management facilities, both the waste is mixed and renders total waste hazardous. About 16,000 million injections are used globally every year. However, most of these are not properly disposed of (Babu et al., 2009). In the United Kingdom, United States, France, Spain and India, the quantity of biomedical waste produced per bed per day are 2.5, 4.5, 2.5, 3.0 and 1.5, respectively (Janagi et al., 2015). Poor biomedical waste management leads to environmental disturbances in terms of air, land and water pollution.

1.2.3. Primary Waste/Agricultural/Lignocellulosic Wastes

The residues from crop production are agricultural waste. The composition of agricultural waste varies with the type and system of agricultural activities. It can be slurries, solids or liquids. It is also termed as 'agro-waste. It consists of animal waste, food processing waste, crop waste, and hazardous waste (pesticides, herbicides, etc.). About 998 million agricultural waste is produced worldwide per year, and 80% is organic (Obi et al., 2016). The fraction of agricultural waste in the total solid waste generation will be increased if the developing countries continue to expand their agricultural system. Earlier, the biomass and agricultural wastes were preferably burnt or converted to organic fertilizers in few cases. Now, these are utilized for energy generation (Mandpe et al., 2021a).

1.2.4. Hazardous Wastes

Any material or substance of chemical, biological or physical nature can pose a potential risk to health, life, and property if hazardous wastes are released in the environment. It is generated from many sources, including wastes to batteries, industrial manufacturing processes, and sludges (Paliya et al., 2019). EPA developed a series of questions in regulatory definition to identify hazardous substances, described below.

(a) Is the material a solid waste? Solid waste means any garbage or refuse, sludge from a water supply treatment plant, wastewater treatment plant, or air pollution control facility and other discarded material from industrial, mining, commercial and agricultural operations, and community activities. It may consist of solid, liquid, semi-solid, or contained gaseous material (Dhote et al., 2020).

(b) Is the waste excluded from the definition of solid waste or hazardous waste? EPA excludes some solid wastes from the definitions ofhazardous waste, and the list is mentioned in the subsection of the 40 Code of Federal Regulations (CFR).

(c) Is the waste a listed or characteristic hazardous waste? There are four lists, the F (wastes from non-specific sources), K (wastes from specific sources), P and U lists, found in title 40 CFR in section 261.

Table 1: The division of F and K list depending on the type of operations. (EPA, 2015).

F-List		K-List	
•	Spent solvent wastes	•	Wood preservation
• Electroplating and other metal finishing wastes		•	Organic chemicals manufacturing
•	Dioxin-bearing wastes	•	Pesticides manufacturing
•	Chlorinated aliphatic hydrocarbons production	•	Petroleum refining
•	Wood preserving wastes	•	Veterinary pharmaceuticals manufacturing
• Petroleum refinery wastewater treatment sludges		•	Inorganic pigment manufacturing
•	Multisource leachate	•	Inorganic chemicals manufacturing
		•	Explosives manufacturing
		•	Iron and steel production
		•	Primary aluminium production

F-List		K-List	
		•	Secondary lead processing
		•	Ink formation
		•	Coking (processing of coal to produce coke)

(d) Is waste Delisted?

The Resource Conservation and Recovery Act (RCRA) issues a delisting process and a process of delisting or removing the generated waste from the list of hazardous waste. The regulations of those mentioned above are given in Title 40 CFR in section 260.22.

1.2.5. Electrical and Electronic Wastes

Waste generating from the end-of-life stage of electrical and electronic equipment (EEE) is termed as 'e-waste'. The e-waste contains components or materials that have economic value if recycled. EEE items comprise of battery, power cord, circuit boards, lead capacitors, activated glass, cathode-ray tubes, and plastic casings. Discarded EEE can be commonly divided into:

- Large household appliances: Washing machines and refrigerators
- Telecom and information technology: laptops, personal computers and monitors
- Consumer equipment: mobile phones, leisure and sporting equipment, televisions, mp3 players)

According to StEP (2013) (Solving the E-waste Problem Initiative), the total e-waste generated in the world in 2012 was 41.36 million metric tonnes. Out of the total generated e-waste in 2012, 10.07 million tonnes and 9.07 million tonnes of e-waste were produced only by China and the United States of America (Perkins et al., 2014). In India, e-waste from cell phones would be increased by 18-fold by 2025 (UNU, 2013). The rapid increase in e-waste is observed due to low recycling rates and short innovation cycles. The consumer life span is continuously decreasing with years. The e-waste generated in developed countries is exported to developing countries under the guise of 'donation' and 'recycling' purposes. The 70% of e-waste final disposal is still unknown or unreported (Ongondo et al., 2011).

1.2.6. Paper and Cardboard

From writing notebooks to posters, cardboard boxes to magazines, papers and cardboards are everywhere. Just by recycling our morning newspaper every day,

41,000 trees can be saved from cutting. For every 100-kilogram municipal solid waste, 40 kilograms of paper are thrown (Wealthy waste, 2011). Paper waste can be classified into three broad categories: mill broke waste, pre-consumer waste and post-consumer waste. The first mill broke waste is generated during the manufacturing processes of paper, which are recycled internally in a paper mill. It mainly consists of trimmings and other paper scraps. Pre-consumer waste is not generated in the paper mill. In contrast, they are discarded before the consumer's use. Lastly, post-consumer waste is discarded after consumers' use and comprises of residential, mixed paper, office paper, old newspapers, old corrugated containers, old telephone directories and old magazines. The paper that is suitable for recycling is known as "scrap paper" and is often utilised for producing moulded pulp packaging.

1.2.7. Other Categories of Wastes

Apart from the mainstream and broadly classified waste in the previous sections, some other categories of waste that hold equal importance when solid waste is the main topic of discussion.

1.2.7.1. Construction and Demolition (C&D) Waste

The waste resulting from the demolition of buildings, roads, or other structures, clearing of land, and construction projects are defined as C&D waste. It comprises of fill materials, plumbing fixtures, carpeting, roofing coverings, land clearing debris, wall coverings, wood, drywall, plaster, *etc.* Every year, in European Union, 450 million tonnes of C&D waste is generated, and in Hong Kong, 38% of the total solid waste generated is C&D waste (Osmani, 2011).

1.2.7.2. Scrap Tires

In the United States, 60% of the rubber is consumed in the manufacturing of tires (Miller, 2010). On average, one scrap tire per capita per year is generated. Non-biodegradable nature and long life make disposal of scrap tires a challenging task. Generally, these are illegally stockpiled, landfilled or burnt. Stockpiling of tires provides grounds for the breeding of vermin, mosquitoes and snakes. Toxic fumes are released for months if they catch accidental or intentional fires. Landfilled tires can damage liners and capping of landfills if it comes to the top of the landfill. They are not ideally meant for landfilling as they occupy the valuable space with 75% void space. The recycling of tires is the only available sustainable option to deal with the waste of scrap tires.

1.2.7.3. Batteries

Batteries play an essential role in daily used products. In the European Union, approximately 725 tonnes of automotive batteries, 172 tonnes of industrial batteries and 145 tonnes of consumer batteries are introduced every year (European Commission, 2019). In the United States, approximately 3 billion dry-cell batteries are purchased per year for laptop computers, portable power tools, toys, power radios, watches and cellular phones and about 99 million wet-cell acid car batteries are introduced to the market every year (Laughlin, 2009). When batteries are improperly disposed of, heavy metals such as mercury, cadmium, nickel and lead are introduced to the environment. Incineration of batteries can cause the release of toxic gases in the air. Recycling of batteries is mainly practised to deal with the used batteries. Moreover, rechargeable batteries are promoted as it may be a substitute for hundreds of single-use batteries.

1.2.7.4. Used Oil

Any synthetic or petroleum-based oil that has been used is termed as used oil. Oils are used for the smooth operation of cars, lawnmowers and many other machines. With time, impurities such as water, chemicals, dirt, metal scrapings are mixed in the oil, and it no longer performs well. For better operation, the used oil is replaced with re-refined or virgin oil. One million gallons of fresh water can be contaminated by used oil from one oil change. Therefore, the handling and disposal of used oil require extra care. Toxic chemicals and heavy metals may be present in the used oil. It is slow to degrade, i.e., persistent in nature and insoluble. It can get stick to everything from birth feathers to beach sand. While changing oil, used oil should be collected in a leak-proof container or can, and spills must be avoided. The same can be dropped at near used oil recycling facilities. The recycled oils can then be again used as a re-refined oil, fuel oils or as a raw material for the petroleum industries.

1.2.7.5. Radioactive and Nuclear Waste

Nuclear power generation produces much less waste as compared to other energy generation facilities. Still, the less generated waste is hazardous and radioactive and must be handled and disposed of carefully. The radioactive waste is produced at each stage of nuclear energy production. Apart from the nuclear plants, radioactive waste is generated from agriculture, manufacturing, research, medicine, minerals exploration and distinctive testing. Any waste that is either intrinsically radioactive or has been contaminated by radioactivity is defined as radioactive waste. The radioactivity decreases with time. The substances with longer half-lived emit alpha and beta, which makes their handling easier. On the other hand, the ones with shorter half-lives emit more penetrating gamma-rays. Eventually, all the radioactive waste decays into a non-radioactive waste.

1.2.7.6. Sanitary Waste

It includes bandages, condoms, nails, used tissues, swept dust, medicines, sanitary napkins, adult diapers contaminated with blood, urine and faeces, used syringes, razors/blades and injection vials. These wastes are generally autoclaved or incinerated. The non-biodegradable raw materials used in the manufacturing of napkins, diapers, *etc.* make sanitary waste a problem.

1.3. Waste Composition in Developing Countries

The 20th century was known as "American Century", while the 21st century is known as "Asian Century" and in the upcoming years, Africa could take the lead. In this era, everyone wants to earn 'as much as possible'. When compared to the developed countries, increased urbanization has been observed more in developing countries. The environmental burdens and waste generation accompanied development worldwide. This unplanned urbanization growth has imposed the increasing demand for municipal health services, sanitation facilities, water and food. Insufficient people and resources limit the developing nations to manage solid waste effectively. Mostly, the developing countries without considering the reality of own room; end up copying the waste management, disposal methods and the legislations of the developed nation. Copying of legislations creates confusions on the ground of roles and responsibilities, clear definition. Furthermore, it causes expenditure constraints for collection, hauling to a disposal site and landfilling. India is one of the largest wastes generating countries in the world. The hike in the gross domestic product (GDP) and increased urbanization followed by high population improved the living standards. The booming population generates more waste. The quantity and complexities of generated solid waste, especially, municipal solid wastes has overburdened the collection and management system. Currently, 39280 tonnes per day of municipal waste is being generated in India (CPCB, 2017). Landfills such as Bordo Poniente in Mexico City, Jardim Gramacho in Rio de Janerio, Sudokwon in Seoul and Laogang in China are amongst the world's largest landfills. More than 10,000 tonnes of waste is landfilled in each (Hoornweg et al., 2013). The developing countries generate waste having a large portion of organic waste and less inorganic waste.

1.3.1. Waste Generation in High Income Groups

High-income groups produce the most waste per capita. The habit of 'use and throw' and a considerable quantity of waste are the aspects, high-income groups are known for. The rich people generate waste four times as a low-income individual does. Highest average annual waste generation is in the high-income groups. The people of this category occupy more space in the form of big houses with more luxury, which results in high waste generation. The fraction of paper, tetra pack, cardboard and metal waste is observed in high percentages among high-income groups as they can afford more processed food, crockery, toys, electric equipment, *etc.* with cardboard and metal packaging (Zia et al., 2017). Shoe fractions and textiles waste are less produced by high-income groups as they transfer it to their maids or low-income families. Income does play an important role in the quantity and composition of waste being generated.

1.3.2. Waste Generation in Low-Income Groups

Low-income groups try to use and reuse the resources up to the maximum possible extent, they end up generating a low quantity of waste. Lowest average annual waste generation is in low-income groups. This is due to their low purchasing capability and money constraints. The waste composition for low-income groups is much different than the high-income groups. It consists more of plastic and glass waste, textile and shoe waste. The standard of living and income affect the quantity, complexity and composition of waste.

1.3.3. Waste Generation in Urban Area

Solid waste is mostly an urban phenomenon. In 1900, 13% of the world's population was urban and produced 300,000 tonnes of waste per day. In 2000, 49% of the world's population became urban and produced 3 million tonnes of waste per day. By 2025, the produced waste would be enough to fill waste collection trucks 5,000 kilometers long every day (Hoornweg et al., 2013). In India, out of the total population, 68% lives in rural areas and 32% in urban areas (Hoornweg et al., 2012). For better opportunities and employment, rural people are shifting to urban areas. According to the World Bank (2003), the urban population in Asia would reach more than 50% of the total population by 2025. This will enforce to expand the urban boundaries of the country. High urbanization causes increased population in urban areas. The urban residents introduce twice as much as waste generated by rural people. In urban areas, the materials discarded from households including the multifamily and single residences, canteens/restaurants, motels and hotels and from the commercial institutes are collected and disposed of by municipal solid waste collection services. Waste composition plays a crucial role in the planning and decision-making stages of waste management (Pratap et al., 2021).

1.3.4. Waste Generation in Rural Areas

In rural areas, there are less manufacturing, low or no food waste and fewer packaged products. The picture is changing, and the waste generation is increasing due to consumerism, increased population and more commercial activities. The land constraint is not a problem in the rural areas for waste management and disposal, unlike urban areas. Lack of awareness and information, rural people are unable to store, manage and dispose of their waste. Cleanliness can be a problem due to improper waste management and handling. Composting followed by the manure application in the agricultural fields has immense scope in rural areas (Mandpe et al., 2019b).

1.4. Waste Composition in Developed Countries

In the United States, an average person generates waste equivalent to their body weight every month. Careful regulations and tracking through well-developed record-keeping systems are utilized for waste management in developed countries. The developed countries generate waste having a large portion of inorganic waste and less organic waste. Waste management is better in developed countries comparatively. There people more concern about the waste they generate and how they handle or manage it. The recycling is practised more often in developed countries due to less economical constraints. Their waste composition is more complex due to high living standard and more luxurious lives (Mandpe et al., 2021b).

1.4.1. Qualitative and Quantitative Analysis of Waste

For the development of an effective waste management system, the foremost step of developing an effective solid waste management system is to create laws, policies, regulations and appropriate administrative practices. The rules can be made as per the requirements which are determined by analyzing the nature of waste. It is important to understand the quantity and composition of waste. The physical and chemical analysis of waste is performed. The qualitative analysis helps in opting the appropriate collection, management, handling, treatment and disposal facility for the waste. Different types of waste require different treatment. Quantitative analysis of waste helps in designing the collection vehicles, treatment plants and landfills (Mandpe et al., 2019a).

1.5. Waste Collection Systems

The waste collection system is the foremost step in waste management. The choice of waste collection system mainly depends on the factors such as choice of treatment, awareness of people, and availability of citizen to pay for the collection, public participation and the cost of collection. Other factors which influence the collection systems are the distance of treatment or disposal area from the collection system and the population of the area. An ideal waste collection system should meet the following criteria.

1. It should meet sanitary regulation in such a way that it could be safe for the environment and human health.
2. The cost should be as low as possible.
3. It should be flexible to answer the timely modification as per the waste generation.
4. The performance and service should satisfy the public need.
5. It should promote schemes like 3R's.

The collection systems are designed based on the factors mentioned above. In this section, the waste collection system is mainly classified into four systems, as mention below.

1.5.1. Containers in the Street, Close to Households

This type of system is used for community services. In this system, the consumer has to bring their waste to the collection bin, which is placed at a specific location in a locality. There is a specific time allotted for the collection of waste. The collection time is generally between midnight to early morning to prevent the traffic jam. The collected waste is transported to the transfer station by small or medium trucks. Due to less or no segregation, the treatment of waste is a challenging task. The cost of installation and maintenance is comparatively low. However, sometimes the overburden of waste in the collection bin lead to nuisance in the area. Therefore, it should be designed by considering the parameters like rate of population in the area, change in the lifestyle which eventually alters the waste generation rate, waste type, easy accessibility, easy to load and unload and withstand external disturbances like rain, storm, wind and other disasters.

1.5.2. Door-to-Door

In door-to-door collection system, all the domestic fractions of solid waste are collected from the residences. The collection vehicles collect the waste from 8 am to 1 pm and unload it to the nearest transfer stations. It is implemented for the area with lower population density. In some areas, the waste collectors charge the waste generators as per the payment per bag or payment per bin.

The advantages of the door-to-door system are:

- Eliminates the waste bins on the streets

- Minimises the anonymously throwing out waste The disadvantages of the door-to-door system are:
- Increases transportation costs
- More unpleasant smell
- Traffic jams on roads
- Cannot be implemented in areas with narrow streets

1.5.3. Pneumatic System

It consists of a series of drop boxes which are located inside houses, in public areas or outdoors, in common areas in apartment blocks, and are connected to a suction pipe via underground pipes. When segregated refuse is dropped into the boxes, it falls in the valves (installed at lower levels) due to the act of gravity. It is stored there temporarily. To collect that stored refuse, there are two systems, including static and mobile.

- Static System: The refuse is emptied regularly, and the collection is coordinated by a central computer.
- Mobile Systems: The chutes are connected to the containers in this system. Suction is created by truck, at specific fixed points; from there different containers are emptied.

The advantages of the pneumatic system are:

- No containers are required for the storage of waste in streets
- Less unpleasant smell
- No collection vehicles and no transportation costs
- Refuse can be dropped at any time of the day
- Waste is segregated

The disadvantages of pneumatic system are:

- High initial investments
- Difficult to implement in areas with a consolidated urban structure
- High operation and maintenance cost
- High energy consumption
- Abrasions can be caused in pipes by glass

1.5.4. Waste Segregation and Transportation

The collected waste is mixed or separated as per the local guidelines. Segregation at source let generators to separate waste into wet (organic waste) and dry (inorganic or recyclables) waste. Different cities have different segregation norms. Segregation, as a source, is mainly practised in developed countries. In developing countries, the waste is not sorted before collection (due to economic constraints) rather recyclables are separated of it at transfer stations or disposal sites. The recyclables sorted from mixed waste may be contaminated. This reduces their market value. After the collection of solid waste from residences, the collection vehicles transport the waste to the nearby transfer stations. The waste is sent to the landfills for the final disposal (Mandpe et al., 2021c).

1.6. Contribution Centres: Drop-Off Centres or Green Points

Drop-off centres or green points are the places where individuals can drop some of the waste (following the sorting norms) which are prone to reuse, recycle, recovery, *etc.* The waste generally accepted in drop-off centres are metals, plastics, cardboard, batteries, electric cells, glass, garden waste, paper, tires, used oils, e-waste. Here, sorting is performed carefully to give specialised treatment to other waste. It reduces the high portion of household items landfilled or incinerated.

1.7. Civic Amenity Site

Another name of the civic amenity site is Household waste recycling centre (HWRC). It is the facility where people can dispose of household waste. They are run by the local authorities. Recyclable waste such as glass, electrotechnical waste, green waste, metals, bulky wastes can be dropped off at these sites. In the United Kingdom, these points are known as tips or dumps.

1.8. Waste Transfer Station

A waste transfer station is an intermediate facility for the temporary storage of municipal solid waste in its journey to landfills or waste to energy facility. At the waste transfer station, a series of steps are carried out with the collected waste from the houses and commercial communities. Unloading of garbage collection vehicles, pre-screening, and removal of inappropriate and hazardous items, such as batteries, with compaction and at last reloading onto large vehicles to their final destination.

1.9. Waste from Markets, Large Stores and Public Utility Service Centres

A large amount of packaging waste leads to a big contribution to the paper and plastic waste. Recycling of packaging waste can eliminate them from entering the landfill premises. The main challenge supermarkets are facing all over the world is food waste. The growing demand for food from costumers leads to high production and supply chain. The various factors (such as including labelling of food, low demand of the product, non-aesthetically pleasing food, high price and replacement of the product by new product) lead food to end up as the piles of waste (Negi et al., 2020).

1.10. Recycling Contamination

Recycling has a major issue known as contamination, affecting the exports, our planet and businesses. In brief, recycling contamination is the introduction of incorrect materials/items in the system when the right materials/items are prepared the wrong way (for instance, recyclables in plastic bags, food residue in containers, shrink wrap recycling mixed in with the cardboard). Recycling contamination can also be termed as aspirational recycling. Even recyclable items, if disposed of improperly, can act as contaminants. A plastic bag into a cardboard stream can act as a contaminant and can cause recycling contamination.

Due to this incorrect disposal, the whole batch of recycled cardboard would be rejected and sent to landfillsand lead to a situation of waste recycling effort!

1.10.1. Types of Recycling Contamination

Contaminants turn recycling into the trash! There are many types of recycling contamination which includes food waste, plastic, and more. Most of the contaminants are easily avoidable, but some of them are worse than others.

(a) Contaminant- Plastic Bags: The worst contaminator of all is the items and plastic bags made from shrink wrap, zip lock bags, bubble wrap, newspaper bags, trash bags, etc.

(b) Contaminant- Food Waste: Remnants of food turn recyclable items into the garbage. The paperboard take-home boxes and recyclable can/ jars full of food that hasn't been rinsed or emptied are the examples of food waste contamination. The paperboard that is employed to carry food usually ends up to landfills due to its potential for recycling contamination.

(c) Contaminant- Loose Shredded-Paper: The long paper fibres in papers make them stand multiple recycling cycles. The shredded paper is not considered as a contaminant as a whole, but when considered loose. It can give rise to many issues which can affect recycling later. At a materials recovery facility, it becomes challenging to separate shredded paper mixed with non-shredded paper. The small size of loose shredded paper causes the recycling issue. It is recommended to keep it in clear bags separately and can be added with their other recyclable items where it would not cause any recycling issues.

(d) Contaminant- Wrong Plastics: The plastics are numbered from #1-7 to separate the recyclables and reusables, accordingly (Seaman, 2012).

Number	Description
1 PETE	Polyethylene terephthalate Soft drink bottles, fruit juice and mineral water containers, cooking oil containers
2 HDPE	High-density polyethylene Milk jugs, laundry detergents, cleaning agents, shampoo bottles, washing and shower soaps, bleaching agents,
3 PVC	Polyvinyl Chloride Fruit, plastic packaging (bubble foil), trays for sweets, food foils to wrap the foodstuff
4 LDPE	Low-density polyethylene Crushed bottles, highly-resistant sacks, most of the wrappings and shopping bags
5 PP	Polypropylene Furniture, consumers, luggage, toys as well as bumpers, lining and external borders of cars
6 PS	Polystyrene Toys, hard packing, refrigerator trays, cosmetic bags, costume jewellery, vending cups, CD cases
7 OTHER	Other Other plastics, including polylactic fibres, nylon, acrylic, fibreglass and polycarbonate

Plastics #1 and #2 are most commonly recycled plastics. Plastics #3 through #7 are recyclables sometimes. Plastics are rejected based on their number and what it contained once

(e) Contaminant- Hazardous Waste: Hazardous waste must be handled as per the rules by EPA and must be cleaned out on a prier basis before it can be recycled.

(f) Contaminant- Bio-Hazardous Waste: Any item containing human fluid such as needles, diapers, sanitary napkins and syringes have the potential to contaminate the whole batch of recycling items!

(g) Contaminant- Unrinsed or Metal-Capped Glass: The excess liquid in the beer or wine bottles can contaminate other items in the recycling and render the whole batch non-recyclable. Metal caps of the bottles are needed to be separated.

(h) Contaminant- Frozen Food Containers: The shiny exterior coating of the boxes that are meant to prevent freezer burn of food prevents the containers from being recyclable.

References

Babu BR, Parande AK, Rajalakshmi R, Suriyakala P, Volga M. Management of biomedical waste in India and other countries: a review. Journal of International Environmental Application & Science. 2009 Jan;4(1):65-78.

Central Pollution Control Board (2017-18), Consolidated annual review report on implementation of solid waste management rules, 2017.

Curbing food wastage in a hungry world. The Hindu BusinessLine, October 30, 2016. https://www.thehindubusinessline.com/opinion/curbing-food-wastage-in-a-hungry-world/article9285737.ece

Dhote, L., Mandpe, A., Paliya, S., Kumar, S., Pandey, R.A., Kumar, R., 2020. Characterization of distillery sludge for its thermal properties and ascertaining its utilization as a low-cost fuel. J. Clean. Prod. doi:10.1016/j.jclepro.2020.120872

European Commission, 2019 April 04. http://ec.europa.eu/environment/waste/batteries/index.htm

Food Wastage in India, And What You Can Do About It. CSR Journal October 31, 2018. https://thecsrjournal.in/food-wastage-in-india-a-serious-concern/

Glass: Material-Specific Data. Environment Protection Agency, 2015. https://www.epa.gov/facts-and-figures-about-materials-waste-and-recycling/glass-material-specific-data

Hoornweg D, Bhada-Tata P. What a waste: a global review of solid waste management. World Bank, Washington, DC; 2012 Mar 15.

Hoornweg D, Bhada-Tata P, Kennedy C. Environment: Waste production must peak this century. Nature News. 2013 Oct 31;502(7473):615.

Implementation of Plastic Waste Management Rules (2016).

Janagi R, Shah J, Maheshwari D. Scenario of Management of Medical Waste in US and UK: a review. Journal of Global Trends in Pharmaceutical Sciences. 2015;6(1):2328-39.

Kumar, S., Negi, S., Mandpe, A., Singh, R.V., Hussain, A., 2018. Rapid composting techniques in Indian context and utilization of black soldier fly for enhanced decomposition of biodegradable wastes - A comprehensive review. J. Environ. Manage. 227, 189–199. doi:10.1016/j.jenvman.2018.08.096

Kumar, S., Bhattacharyya, J.K., Vaidya, A.N., Chakrabarti, T., Devotta, S., Akolkar, A.B., 2009. Assessment of the status of municipal solid waste management in metro cities, state capitals, class I cities, and class II towns in India: An insight. Waste Manag. 29, 883–895. https://doi.org/10.1016/j.wasman.2008.04.011

Kumar, S., Smith, S.R., Fowler, G., Velis, C., Kumar, S.J., Arya, S., Rena, Kumar, R., Cheeseman, C., 2017. Challenges and opportunities associated with waste management in India. R. Soc. Open Sci. 4. https://doi.org/10.1098/rsos.160764

Laughlin RB. Environmental Protection Agency-Battery Waste. 2009 Nov 25. http://www.epa.gov/waste/conserve/materials/battery.htm

Mandpe, A., Kumar, S., Lakshmikanthan, P., Hettiarachchi, H., 2019a. Mining for recovery as an option for dumpsite rehabilitation: Case study from Nagpur, India. J. Environ. Eng. Sci. 15, 52–60. doi:10.1680/jenes.19.00021

Mandpe, A., Paliya, S., Kumar, S., Kumar, R., 2019b. Fly ash as an additive for enhancing microbial and enzymatic activities in in-vessel composting of organic wastes. Bioresour. Technol. 293, 122047. doi:10.1016/j.biortech.2019.122047

Mandpe, A., Kumari, S., Kumar, S., 2020a. Composting: A Sustainable Route for Processing of Biodegradable Waste in India. Org. Waste Compost. through Nexus Think. 39–60. doi:10.1007/978-3-030-36283-6_3

Mandpe, A., Tyagi, L., Paliya, S., Chaudhry, S., Motghare, A., Kumar, S., 2021a. Rapid-in-house Composting of Organic Solid Wastes with Fly Ash Supplementation: Performance Evaluation at Thermophilic Exposures. Bioresour. Technol. 337, 125386.

Mandpe, A., Yadav, N., Paliya, S., Tyagi, L., Ram Yadav, B., Singh, L., Kumar, S., Kumar, R., 2021b. Exploring the synergic effect of fly ash and garbage enzymes on biotransformation of organic wastes in in-vessel composting system. Bioresour. Technol. 322, 124557. doi:10.1016/j.biortech.2020.124557

Mandpe, A., Pratap, V., Lakshmikanthan P., 2021c. Landfill mining. In Current Developments in Biotechnology and Bioengineering (pp. 197-229). Elsevier. doi: 10.1016/B978-0-12-821009-3.00010-5

Manyapu, V., Mandpe, A., Kumar, S., 2019. Corrigendum to "Synergistic effect of fly ash in in-vessel composting of biomass and kitchen waste" [Bioresour. Technol. 251 (2018) 114–120](S0960852417321740)(10.1016/j.biortech.2017.12.039). Bioresour. Technol. 277, 225. doi:10.1016/j.biortech.2018.12.031

Miller C. Waste360. 2010 Mar 01. https://www.waste360.com/Recycling_And_Processing/scrap-tires-201003

Negi, S., Mandpe, A., Hussain, A., Kumar, S., 2020. Collegial effect of maggots larvae and garbage enzyme in rapid composting of food waste with wheat straw or biomass waste. J. Clean. Prod. 258. doi:10.1016/j.jclepro.2020.120854

Obi FO, Ugwuishiwu BO, Nwakaire JN. Agricultural waste concept, generation, utilisation and management. Nigerian Journal of Technology. 2016;35(4):957-64.

Packaging waste statistics, January 2019. https://ec.europa.eu/eurostat/statistics-explained/index.php?title=Packaging_waste_statistics

Paliya, S., Mandpe, A., Kumar, S., Kumar, M.S., 2019. Enhanced nodulation and higher germination using sludge ash as a carrier for biofertilizer production. J. Environ. Manage. 250, 109523. doi:10.1016/j.jenvman.2019.109523

Perkins DN, Drisse MN, Nxele T, Sly PD. E-waste: a global hazard. Annals of global health. 2014 Jul 1;80(4):286-95.

Pratap, V., Bombaywala, S., Mandpe, A., Khan, S., U., 2021. Solid Waste Treatment: Technological Advancements and Challenges. In Soft Computing Techniques in Solid Waste and Wastewater Management (pp. 215-231). Elsevier. https://doi.org/10.1016/B978-0-12-824463-0.00014-8

Seaman G. Plastic by the Numbers. 2012 May 2. Retrieved from: https://learn.eartheasy.com/articles/plastics-by-the-numbers/

StEP Initiative. StEP Annual Report. 2013. Available at: http://step-initiative.org/tl_files/step/StEP_AR/StEP_AR.html. Accessed October 14, 2014.

Fact Sheet on Plastic Waste in India. TERI 2018. https://www.teriin.org/sites/default/files/files/factsheet.pdf

United Nations University (UNU), Solve the E-waste Problem (StEP), Massachusetts Institute of Technology (MIT), National Center for Electronics Recycling (NCER). World e-waste map reveals national volumes, international flows. 2013. Available at: https://www.vie.unu.edu/file/get/11505.pdf. Accessed October 14, 2014.

Miller C. Profiles in Garbage: Yard Waste. January 21, 2013. https://www.waste360.com/research-and-statistics/profiles-garbage-yard-waste

Ongondo FO, Williams ID, Cherrett TJ. How are WEEE doing? A global review of the management of electrical and electronic wastes. Waste management. 2011 Apr 1;31(4):714-30.

Osmani M. Construction waste. InWaste 2011 Jan 1 (pp. 207-218). Academic Press. Recycling Paper from Paper Waste. Wealthy Waste. November 9, 2011.

Wainaina, S., Awasthi, M.K., Sarsaiya, S., Chen, H., Singh, E., Kumar, A., Ravindran, B., Awasthi, S.K., Liu, T., Duan, Y., Kumar, S., Zhang, Z., Taherzadeh, M.J., 2020. Resource recovery and circular economy from organic solid waste using aerobic and anaerobic digestion technologies. Bioresour. Technol. 301, 122778. https://doi.org/10.1016/j.biortech.2020.122778

World Bank Annual Report. 2003. Yard Waste. Environmental Encyclopedia, 2003. https://www.encyclopedia.com/environment/ encyclopedias-almanacs-transcripts -and- maps/yard-waste

Zia A, Batool S, Chauhdry M, Munir S. Influence of Income Level and Seasons on Quantity and Composition of Municipal Solid Waste: A Case Study of the Capital City of Pakistan. Sustainability. 2017 Sep 6;9(9):1568.

2

Greenhouse Gases Emissions from Agriculture: Management and Mitigation Strategies

Jaipal[1], Hardeep Rai Sharma[2] and Asha Gupta[3]*

[1]*Department of Civil Engineering, DAV College of Engineering and Technology Kanina, Mohindergarh-123027, Haryana, India*
[2]*Institute of Environmental Studies, Kurukshetra University Kurukshetra-136119, Haryana, India*
[3]*Department of Environmental Science and Engineering, Guru Jambheshwar University of Science and Technology, Hisar-125001, Haryana, India*

ABSTRACT

Greenhouse gases (GHGs) cause a greenhouse effect by absorbing and emitting radiant energy within range of the thermal infrared radiation. By this greenhouse effect current temperature of the earth is around 59°F otherwise it would be around minus 2°F. Agricultural activities serve as both sources and sinks for GHGs. Agricultural land has a capacity of carbon sequestration which eliminate CO_2 from the atmosphere. There are the various carbon sinks on the planet in which soil is the largest terrestrial sink. Mass production of livestock and use of manure and fertilizer is the primary source of GHGs in the agriculture. Deforestation and soil degradation caused due to agricultural expansion emit GHGs like carbon dioxide. The harmful impact of agriculture on the environment will increase with time; especially with the continuous increase in the world population, which in turn raises the danger of food scarcity. However, by utilizing agricultural practices that require negligible disturbance to the soil and stimulate carbon sequestration, farmers feasibly capable to slow or even reverse the carbon loss from their fields. This chapter will discuss various aspects of GHGs like emissions, trends, impact and control related to agriculture in the Indian context.

Keyword: Greenhouse gases, greenhouse effect, thermal radiation, carbon sequestration, deforestation, food security

1.1 Introduction

Increased emission of greenhouse gases results in climate change which becomes one of the most prominent environmental issues to the world today. Carbon dioxide (CO_2) is the major greenhouse gas whereas nitrous oxide (N_2O) and methane (CH_4) are other substantial contributors to climate change (IPCC, 2014, Tongwane et al., 2016). However, these gases are emitted in minimal quantities in comparison to carbon dioxide but impart more global warming effect by absorbing more infrared radiations (Oertel et al., 2016). The warming potential of one kg of CH_4 is 23 times more than one kg of CO_2 while in the case of N_2O, it was nearly 300 times more than CO_2 (Ramaswamy et al., 2001, Bharti et al., 2019; Biswas et al., 2010). Around one-third of methane and two –thirds of nitrous oxide come from soil (Prather et al., 1995, Tubiello et al., 2015, Smith, 2016). As a result of global warming, oceans are going up; snow and ice are melting fast, and the sea level is increasing. Moreover, the considerable impact of global climate change had also been observed on crops, livestock, soil, fisheries and pests (Schleussner, 2016, Dhar et al., 2017).

Nation's most fundamental obligation, *i.e.,* food, nutritional and livelihood security, is served through agriculture. It provides gainful employment to around two-thirds of the workforce, which accounts a remarkable share in India's financial system. Several industries depend on agriculture for their raw materials. In recent years, despite making significant progress, Indian agriculture is facing big challenges of:

- Stagnation of total area sown
- Stabilizing of yield levels
- Decline of soil health
- Depletion in *per* person land availability
- Climate change

As a result, the task of providing food and nutritional security to around 1619 million people by 2050 will become a challenging issue because > 80% farmers are small (having 1-2 hectare) and marginal (having agricultural land less than one hectare) landholders with inadequate cropping capacity. However, the international community is involved in making policy to bring down the discharge of greenhouse gases and reduce the threat of climate change under the directions of United Nations Framework Convention on Climate Change (UNFCCC) (UNFCCC, 2008).

Comparative to other sector agriculture is at high risk of climate change. India's stake in climate change prescribes that adaptation is established in agricultural policy of our country. This agricultural policy would progressively require to conforming with the climate-associated objectives in such a way that is slightly harming to the other basic objectives of the policy. Agriculture not only acts as big source for emission of greenhouse gases but also provide robust solution to the problems (IPCC, 2014).

There is a requirement for careful assessment of mitigation approaches especially related to agriculture. Such assessment gives an idea that how much adopted mitigation approaches are compatible with the objectives of farmer's livelihood and food security and how agriculture can contribute to the abatement of greenhouse gases (Frank et al., 2017, Wainaina et al., 2020).

2.2 Objectives of this Chapter are as Follows

i. To assess the emission of greenhouse gases from agriculture in India

ii. To assess the trends and drivers of greenhouse gas emission

iii. To understand the potential and economic feasibilities of various mitigation options

iv. To recognize how and in which way climate change can affect agriculture in India

v. To enumerate the farm applications which can help in the carbon sequestration and emissions control

vi. To lay out the blueprints for an agricultural strategy that will oversee the climate related issues un- compromising the general objectives of food and livelihood security

2.3 Sources for the Emission of Greenhouse Gases in the Sector of Agriculture

Primarily the greenhouse gases such as CO_2, CH_4 and N_2O are produced through various agricultural activities and contribute global warming (Table 1).

Table 1: Sources of greenhouse gases

CO_2	N_2O	CH_4
The main source for CO_2 emission in agriculture sector is the soil management application like tillage. In such type of application surface area of the soil has increased due to the breakdown of aggregates of the soil. In such environment organic matter of the soil easily undergoes biological decomposition and generates CO_2. • Burning of crop residues and use of conventional	Potential source for emission of N_2O in the sector of agriculture are: -Cultivation of nitrogen fixing crops • Cultivation of soil with high organic content • Holding of crop residue • Application of chemical fertilizer and manure/ other organic material. The other sources which indirectly affect the emission of N_2O are:	Potential source for the production of methane in agriculture are the rice fields. • Emission of methane has increased by the application of organic manure and high content of organic carbon in puddle soil.
source of energy in several agricultural operations	-Application of tillage, drainage, irrigation, following of fields. Emission of N_2O has increased by leaching, surface runoff, subsequent atmospheric deposition and volatilization of applied nitrogen (Mosier et al., 1998,	-Residue of crop release
-Manufacturing of farm implements, fertilizers and		methane on its burning
pesticides (Janzen, 2004, Smith, 2004, Sejian et al,		-Ruminants (Enteric
2016, Ren et al., 2017).		fermentation) (Yan et al.,
	Oenema et al., 2005, Phillips et al., 2009, Linquist, et al.,	2003).
	2012, Wang et al., 2018).	

Sources of greenhouse gases

The global emission of greenhouse gases was estimated around 49.4 Gt CO_2 eq. in the year, 2016, in which India contributes about 6.7 % of total emission (3.3 Gt CO_2 eq.) (WRI CAIT, 2017, FAO, 2018) Contribution of greenhouse gas emission concerning agriculture was observed around 11.8% at the global level of total GHGs emission. 7% of global emission from agriculture was contributed by Indian agriculture. Energy and agriculture sector are two main contributors to the greenhouse gas emissions at national as well as global level. Sector-wise contribution to greenhouse gas emissions is illustrated in Fig.1 and Fig.2 (FAI, 2017, IEA, 2018).

At the global level, China is the biggest emitter of greenhouse gases which emitted around 26% greenhouse gases followed by United States (13%), European Union (7.8%) and India (6.7%) (Liu et al., 2019). At global level India is observed as a 2nd largest emitter of greenhouse gases in the sector of agricultural while in all the sectors India is observed as 4th largest emitter. At the National level, an average emission (from 1990 to 2016) of greenhouse gases from different subsector of agriculture is presented in Fig.3 (WRI CAIT, 2017, Abbas et al., 2020). Among these subsectors, enteric fermentation is the main contributor which contributes around 39.3% followed by manure left on pasture (15.2%), synthetic fertilizers (11.8%), rice cultivation (10.1%), manure management (6.8%) and others such as burning savanna, crop residue, manure applied to the soil, cultivation of organic soil, burning of crop residue (16.8%). Among the various rice ecosystem, maximum 34% greenhouse gas was emitted by irrigated regularly flooded rice after that rain-fed flood, drought-prone, deep water, and irrigated multiple aerations aided around 18%, 16%, 8% and 6% of methane, respectively (Bhatia et al., 2012, Bhatia et al., 2013, Pathak et al., 2014). In different states of India, its emission varies from 4 to 190 kg *per* hectare. In the case of nitrous oxide, among the various sources, fertilizers are the major contributors which emit around 77% of total N_2O emission. In India, its emission varies from 0.18 to 9.11 kg *per* hectare. Indian croplands shared around 47 % of agricultural land and emits 61% of direct N_2O emissions while grasslands shared around 53 % of agricultural land and emits 39% of N_2O which indicates that cropland constitute 6% lesser area of agricultural land and emits 22% more N_2O than grassland.

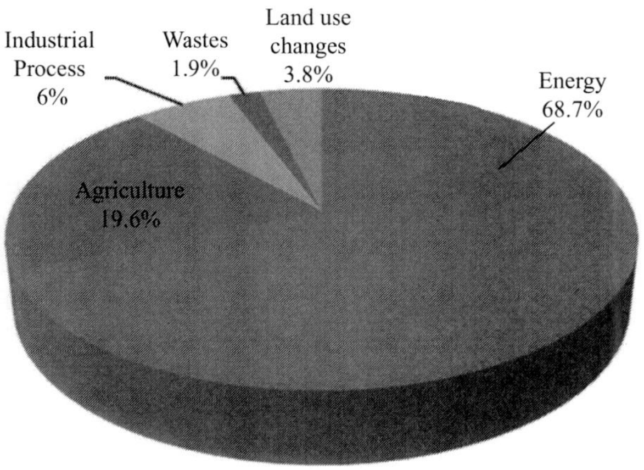

Fig. 1: Emission of greenhouse gas from various sectors in India during 2016

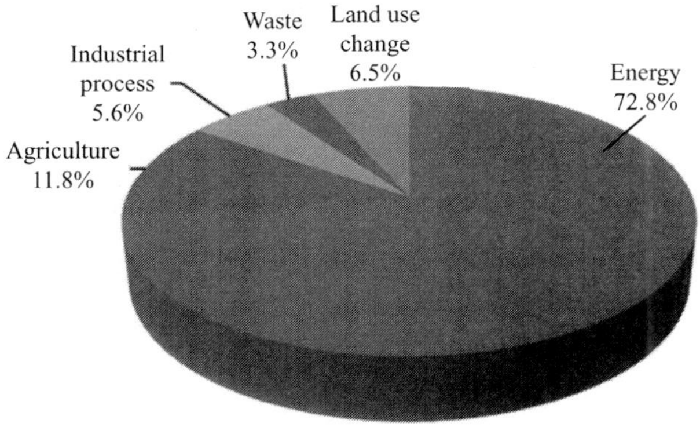

Fig. 2: Emission of greenhouse gas from various sectors in World during 2016

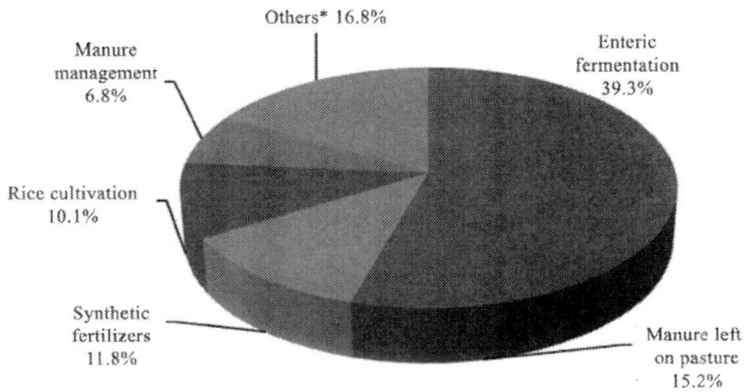

Fig. 3: Average greenhouse gas emission from agriculture sector in India during 1990 to 2016

*(Burning Savanna, crop residue, manure applied to soil, cultivation of organic soil, burning of crop residue)

2.4 Agriculture and the UNFCCC

The world's population is growing at a very alarming rate which has crossed 7 billion and probably to attain 8 billion by 2030. While limiting the GHGs emission, to fulfil the food and energy security of this growing population is a global challenge. UNFCCC is working with the objectives to stabilize the concentrations of GHGs in the atmosphere. Concerning climate changes, the objective of UNFCCC negotiations is mainly focused on the issue of whether mitigation or adaptation should be employed in global agriculture. The views of some group of parties, mostly the developing countries supported priority to adaptation by the following facts:

- In developing countries, agriculture is being highly diverse, unorganized and heterogeneous, which creates difficulties in mitigation of greenhouse gases emissions through an organized work program.
- Agriculture needs to be intensified for ensuring food security that may enhance GHGs emission in the short-term.
- The issue of common but differentiated responsibilities and non-availability of cost-effective and straightforward mitigation technologies.

However, the advanced countries contended that agriculture could be a possible sink for GHGs without any commitment with the increasing global food demand. Although, a lot of scopes for adaptation with co-benefits for mitigation in agriculture has been acknowledged by both the parties. Production of fuel, feed and food can be enhanced by growing of stress-tolerant breeds/ cultivars of crops, forest trees fish and livestock without increasing emission of GHGs. Moreover, through bio-energy plantation, degraded land restoration and grazing land management, a massive potential for GHGs mitigation can be accomplish by sequestration of carbon in the soil.

2.4.1 Negotiations on Agriculture in the UNFCCC

Various protocols/ treaties/conventions were held concerning negotiations on agriculture in which a major step was taken by UNFCCC in reducing GHGs emission from various sources in the Kyoto Protocol (1997); however, the reduction of GHGs emission from agriculture in this protocol is optional. Other conventions are:

13th Conference of the parties (COP 13), Bali, Indonesia, 2007: Under this convention a subsidiary body as an Ad hoc Working Group on Long-term Cooperative Action (AWG-LCA) was formed via Bali Action Plan to regulate a comprehensive process to enable the fruitful and continual execution of the convention. Moreover, agriculture has been considered as a subject of discussion. In April 2008 at Bonn, Germany, it invited its Chair to prepare a report by collecting the proposals presented by organizations such as the Food and Agriculture Organization (FAO). Accordingly, a proposal for inclusion of GHGs mitigation through carbon sequestration in soil was submitted by Food and Agriculture Organization (UNFCCC, 2008).

COP14, Poznan, Poland 2008: In the year 2008, **COP14** document was updated by AWG-LCA chair on the bases of the submissions received from various observer organizations and their views on agriculture were considered in the preparation of documents.

COP15, Copenhagen, Denmark 2009: Under this COP in respect of agricultural sector, it has been decided that all the Parties will consider their regional and national priorities. The main objectives of this COP were to promote control process and technological transfer to reduce the emission of greenhouse gases from the anthropogenic sources, to support adaptation for severe effect of climate change and to improve agricultural production in sustainable way for providing livelihood and food security. Therefore, in this conference food security and climate change related issues were taken into consideration.

COP16, Cancun, Mexico 2010: In **COP16** text report of Ad.Hoc.Working. Group.on.Long-Term Cooperative.Action (AWG-LCA) meeting held at Tianjin, China in October 2010 on agriculture and climate change was presented.

COP17, Durban, South Africa2011: In **COP17,** Subsidiary Body for Scientific and Technological Advice of UNFCCC was requested to consider agriculture-related issues for exchanging views and adopting a decision in this regard at the next COP.

COP18, Doha, Qatar 2012: Subsidiary Body for Scientific and Technological Advice concluded meeting along with draft decision that the matter should be referred to COP19 for further consideration.

COP19, Warsaw, Poland 2013: COP 19 was held in November 2013 at Warsaw, Poland in which agriculture-related issues were further discussed for an appropriate decision.

The issues related to agriculture were again discussed in this COP19 for a decision. For this, an in-session workshop was conducted in which several parties presented their views on issues related to agriculture. The UNFCCC prepared a report to discuss the matter in its forthcoming session. The world community must take the decision carefully to benefit the maximum number of farmers (small and marginal) in developing and poorly-developed countries for their contribution in global mitigation efforts by the use of climate-smart agriculture, climate financing incentives and carbon markets.

COP 20, Lima, Peru, 2014: This COP provided a forum for exploring a climate justice approach on climate-smart agriculture and food security. It also prepared a draft report for establishing an international climate agreement during COP 21 (De-Pinto et al., 2020).

COP 21, Paris, France, 2015: An international agreement was made to set a target for limiting the global warming below 2°C as compared to pre-industrial level and agriculture sector was kept as the key sector in addressing the climate change. Process and techniques for low greenhouse gas agriculture were discussed. In this COP, all the participating nations were agreed to cut back greenhouse gas emission, and ultimately a universal agreement on climate change action was created (UNFCCC, 2015).

COP 22, Marrakech, Morocco, 2016: It was focused on turning the landmark Paris Agreement on climate change into a detailed blueprint for action. For ensuring the linkage between agriculture, food security and climate change, this COP provided an opportunity to Food and Agriculture Organization to give technical advice to support the involvements of member countries in the negotiations and in meeting their UNFCCC commitments.

COP 23, Bonn, Germany, 2017: In this COP a Koronivia joint work on agriculture was developed for emphasizing the importance of food security and agriculture in the agenda of climate change and member countries agreed to work jointly for making sure that agricultural development ensures both a reduction in emissions and increased food security in respect of climate change (UNFCCC, 2018).

COP 24, Katowice, Poland, 2018: During this COP, practices and techniques for making productive and resilient agricultural systems while reducing emissions were discussed by scientists, industry representatives and farmers of member countries. It highlighted the best-proven practices and realistic possibilities with real-world examples for strengthening the agricultural resilience, mitigating agricultural greenhouse gas emission and increasing agricultural productivity.

COP 25, Madrid, Spain, 2019: COP 25 played an important role in bringing the Paris Agreement into the force and developed the guidelines for the working of the international carbon market.

2.4.2 Tools for Agricultural GHGs Monitoring, Reporting and Verification

Low confidence and limited awareness in the monitoring of agricultural interventions inhibited the inclusion of agriculture in the emission offset markets and climate change policy even after the significant mitigation potential of various agricultural practices. Monitoring, reporting and verification (MRV) system are required to provide incentives for adopting GHGs emissions mitigation approaches in agricultural practices towards climate-smart agriculture. The important aspects of MRV are:

- Technical measurement of agricultural emissions and their removal.
- Quality of the data collection.
- Adoption of the evaluation process.
- Methodologies for estimation of emission reductions.

The developments in modelling tools, sampling techniques for field measurements and remote sensing technology strengthened the capability of agricultural GHGs projects to meet several challenges along with the rising standards of credibility and rigour in the emissions offset markets.

A technique such as remote sensing can be used for recording the land-cover and land use change over time and for tracking the changes in the relative distribution of land-use classes. Remote sensing data, along with field measurements, can be integrated into models and then estimates with various levels of perfections will be produced on the bases of existing data and local ecology. However, information such as soil characteristics, growth rates, biomass pools, decomposition rates, land-use change, climate, current and historic natural disturbance, management practices and conversion equations and estimates of variability and error can be used as the typical inputs for the model. For the model in the tropical region, reliable input data remain a major obstacle.

2.5. Trends and Drivers of GHGs Emission in Indian Agriculture

Trends and drivers of greenhouse gases in the agricultural sector are as under: India is the world's 2nd largest emitter of greenhouse gases in the sector of agriculture. At National level, greenhouse gas emission has increased from 406 Mt CO_2 eq. to 700 Mt CO_2 eq. during the year 2010 to 2016.

2.5.1 Trends in GHGs emission

In the sector of agriculture, India contributed 19.6% of country's total greenhouse gas emissions. Current greenhouse gas emission is observed around 700 Mt CO_2 eq. while it was around 406 Mt CO_2 eq. during the year 2010 (Sapkota et al., 2019). Enteric fermentation, rice cultivation and manure management were observed the major contributors to the methane emission. Still, applications of fertilizers and manure were the major contributors in the case of nitrous oxide (IPCC, 2014). Compared to other greenhouse gases, N_2O emission is of greater environmental concern because of its long residence time of 114 years in atmosphere, higher (265 times) global warming potential than that of carbon dioxide on a 100-year time scale, which make it 3rd most abundant greenhouse gas after CO_2 and CH_4 (IPCC, 2014, MoEFCC, 2018, Fagodia et al., 2020). During the year 2010 nitrous oxide emission was increased by 73% while the methane emission was raised by 18%. It was because nitrogen consumption has increased by 66% times in the country since 1960. The maximum use of chemical fertilizers along with other agri-inputs and an increase in the population of lives stock act as major drives which increase the GHGs emissions. However, the relative contribution of Indian agriculture to total GHGs emission from all the sectors has decreased from 33% to 15% during 1970 to 2016 while the contribution to global agriculture was increased from 9% to 10% during this period (Pathak et al., 2014, Fagodia et al., 2020).

2.5.2 Trends in Emission Intensity

In India, the emission intensity of greenhouse gases have decreased by 18% *per* ton of food production during 1970 to 2016 mainly because of increase in agricultural production of the country by using high-yielding crops varieties, animal breeds along with better animal and crop management practices without a considerable increase in significant sources of methane such as ruminant population and area under rice (Olivier et al., 2017). However, greenhouse gas emission *per* hectare of land for India is increased by 88% during this period. The greenhouse gas emission from agriculture is projected to increase by 19% up to 2030 as compared to the current level. As per the Kyoto protocol, the emission of greenhouse gases should be at the level of 1990 means it should be reduced by 21% up to 2030.

2.5.3 Drivers of GHGs Emission

Major drivers for GHGs emission and agricultural change are: food and energy security, increase in population, increase in *per* capita income, developments in technologies, cost of carbon markets in the world, shortage of water and labour, liberalization and globalization, changes in dietary trends and food consumption, synergy with sustainable development and co-benefits of GHGs mitigations (Hazell and Wood, 2008, Gibbs et al., 2010, Harvie et al., 2011, FAO, 2017). Such type of changes can impart alarming pressure on agriculture which can result in considerable changes in greenhouse gases emission. These drivers will work at the local, national and global level.

2.6. Mitigating GHGs Emission from Indian Agriculture

Agriculture due to its unique property offers the mitigation opportunities such as improving manure management; production and management of nitrogen fertilizer; sustainable intensification and reducing emission from rice cultivation and enteric fermentation from supply-side also offer the mitigation opportunities such as carbon sequestration in agricultural systems (Paliya et al., 2019); shifting dietary trends and reducing food wastes from the demand side. According to Intergovernmental Panel on Climate Change (IPCC) (IPCC, 2014) the mitigation options in agriculture are i) Nutrient management (urease/nitrification inhibitor and type of fertilizer) ii) Plant management (improved varieties rotation, cropping and rotation system) iii) Water management (improved drainage, water application) iv) Residue/tillage management (reduced tillage, residue retention) v) Bio char application vi) Land-use changes (bio energy crops, agro forestry) (Pathak et al., 2015).

The managed soils are observed as the primary source for CH_4 and N_2O; both have significant global warming potential ranging from 240 kg CO_2 eq. ha-1 in pulses to 3700 kg CO_2 eq. ha-1 in continuously flooded rice (Gupta et al., 2009). The global warming potential of some other crops such as wheat, rice (intermittent flooding), pulses, maize, oilseeds and vegetables under the application of fertilizers is expressed in Table-2.

Table 2: Global warming potential of crops in India

Crops	Application of fertilizer (Kg ha-1)	Global warming potential (kg CO_2 ha-1)
Wheat	120-150	340-450
Rice (intermittent flooded)	120	900-1050
Pulses	40-60	180-240
Maize	120	320-365
Oilseeds	60-80	220-275
Vegetables	150-300	440-75

Sources: Abbas et al., 2020, Pathak et al., 2015, SNS, 2012

Agriculture also itself has the cost-effective mitigating potential for GHGs emission via the implementation of low carbon in agricultural technologies and best management practices (Abbas et al., 2020). Indian agriculture has potential to mitigate 85.5 Mt CO_2 eq. *per* year on the adoption of integrated agronomic practices, energy-efficient technology, improved water and nutrient management practices, improved fertilizer production technology, zero tillage, improved diet in animals, green fodder supplements, soil salinity reclamation techniques and restoring degraded land, of which 80% is delivered by the cost-effective option using best management practices (Sapkota et al., 2019, ICAR-CSSRI, 2016, Miller et al., 2018). Mainly, three approaches for mitigation of GHGs are used:

(i) **Reduction in emissions:** The emissions of greenhouse gases such as CO_2, CH_4, and N_2O can be decreased through appropriate management of nitrogen and carbon flows in the agricultural ecosystems. For instance, methane emission can be suppressed through managing livestock which makes the most efficient use of feeds (Clemens and Ahlgrimm, 2001) and nitrous oxide emission can be reduced through the practices that deliver added- N to the crops (Bouwman, 2001, Feliciano et al., 2017). The approaches used in the reduction of GHGs emission vary from region to region as they depend on the local conditions.

(ii) **Enhancement of removal:** Agricultural ecosystem stored the large carbon reserve in soil organic matter (IPCC, 2001), and these systems lost the significant amount of carbon, some of which can be improved via better management practices (Paustian et al., 1998). By using various practices which suites to local condition; a significant portion of soil carbon can be stored.

Some practices increase the photosynthetic input of carbon and decrease the loss of stored carbon via erosion, fire, and respiration, thereby enhancing the carbon reserves (Lal, 2004). By planting the perennial plants on agricultural lands, significant amounts of vegetative carbon can be fixed (Albrecht and Kandji, 2003). The agricultural lands by oxidation process can also remove CH_4 from the atmosphere (Smith and Conen, 2004).

(iii) **Avoiding emissions**: The emission of CO_2 can be avoided by using such type of management practices which project the cultivation of new lands that are under grassland, forest, and non-agricultural vegetation (Foley et al., 2005). GHGs release from agriculture can also be avoided by sequestering carbon in soil and decreasing emissions of CH_4 and N_2O from the soil via changes in land-use management practices. The use of perennial plants having extensive root systems increases the amount of carbon stored in the soil. Residues of cultivation systems especially reduce deep tillage and encourage the build-up of soil carbon. The emission of nitrous oxide and methane can be reduced through proper management of fertilizer use, irrigation, soils and changes in crop genetics (Cassman et al., 2003, Galloway et al., 2003). Such alternatives not only mitigate global warming but also improve soil fertility. The emission of GHGs can be reduced by using the agricultural feedstock such as crop residues, dung and energy crops as the substitutes of fossil fuels in energy production (Cannell, 2003, Schneider and McCarl, 2003). By developing policies and incentives, farmers may be motivated to harness the profits of improved soil health and water and energy-use efficiencies by adopting these mitigation options. Various possible required interventions, key challenges, significant benefits for encouraging low-carbon and cost-effective GHGs mitigation technologies for wheat and rice crops are given in Table-3.

Table 3: Key challenges and interventions required for improvement of mitigation potential of technologies used in the mitigation of greenhouse gases evolved in wheat and rice crops (Indo-Gangetic plains)

Technology	Major advantages	Major challenges	Interventions required for improvement of mitigation potential of technologies
Wheat			
Nitrification inhibitor and site-specific nitrogen management	• Increase in nitrogen use efficiency • Reduction in nitrogen losses	• Non-availability in market • The high technical application process • The high cost of production	• Provision of incentives to the farmers • Introduction of carbon credits for mitigation • Generation of public awareness • The motivation of industries to manufacture these products
Zero tillage	• 70-90 L saving in diesel • Saving in water	• Lack of awareness • Non-availability and maintenance of quality zero-till drill • Knowledge-intensive	• Refinement of technology • For operation and maintenance of drill, skill development is required • Generation of awareness
Integrated nutrient management/organic farming	• Reduction GWP • Increase in income • Improvement in soil health • Increase in nitrogen use efficiency	• Poor quality of manure • Non-availability of manure • High transportation cost	• Encouragement of biogas plants for energy and manure use • Developments of carbon credits for biogas • Promotion of agri-wastes for compost making
Rice			
Drip and sprinkler irrigation	• 25-30% saving in irrigation water • Saving in energy • Higher water and nutrient use regulations	• Practical skill is required • Need for infrastructure for installation and maintenance • High initial cost	• Initiation of carbon credits for mitigation • Generation of public awareness • Organization of training for skill development • Introduction of subsidy for installation

nitrogen management/ leaf colour chart (LCC) for the specific sites	• 10-15% increase in nitrogen use efficiency • Reduction in nitrogen losses	• Inadequate soil testing facilities • Lack of awareness	• Provision of incentives to the farmers •Introduction of carbon credits for mitigation • Generation of public awareness • Establishment of more soil testing laboratories
Nitrification inhibitor/urea super granules	• 10-15% increase in nitrogen use efficiency • Reduction in nitrogen losses	• Technical application process • Non-availability in market • High production cost	• Provision of incentives to the farmers • Introduction of carbon credits for mitigation • Generation of public awareness • The motivation of industries to manufacture these products
Alternate wetting drying/ Mid-season drainage	• 25-30% saving in irrigation water • 30-35% reduction in methane emission • 10-15% saving in labour	• Weed problems • Difficulties and risk in rain fed area • 25-30% increase in nitrous oxide emission • 5-10% yield loss	• Accurate weather forecasting • Incentive provisions for conserving water • Introduction of carbon credits for mitigation • Development of environment-friendly herbicides • Development of irrigation facilities
Crop diversification	• 50-60% saving in irrigation water • 90-100% reduction in methane emission • 20-25% saving in labour	• Lack of awareness • Limited post-harvest facilities • Inadequate market • 45-50% increase in nitrous oxide emission	• Development of post-harvest facility • Development of market infrastructure • Awareness generation • Provision of the minimum spot price for alternate crops

Sources: Pathak et al., 2009 and 2014, Sapkota et al., 2017, Lungerska and Chakir, 2018, Sharma et al., 2018

2.6.1 Mitigation of CH_4 Emissions

The potential source for methane is rice fields submerged underwater, microbial anaerobic degradation of organic matter and use of organic matter in puddle soil. Emission of methane from rice cultivation can be mitigated by adopting strategies such as direct seeding of rice (DSR) and rice intensification system (RIS); changing water management system, especially alternate wetting and drying (AWD); improved organic matter management system involving aerobic decomposition via composting (Kumar et al., 2018); incorporation of composted organic matter into the soil during off-season drained period; rice cultivars along with some unproductive tillers, high harvest index, high root oxidative activity; application of fermented manure instead of unfermented manure and intermittent flooding approach (Pathak et al., 2014, Zhao et al., 2011).

Direct seeding of rice (DSR) and rice intensification system (RIS) can be used as the potential option in the reduction of methane emission as they do not need continuous soil submergence as required in case of conventional puddle transplanted rice. As compared to traditional puddle transplanted rice, DSR and RIS can reduce the GWP by 35-75% (Pathak et al., 2013). But the intermittent flooding system can reduce only 25-30% over continuous flooding system.

Alternate wetting and drying (AWD) practice initially involve flooding a field and then drying its upper layer in repeated cycles throughout the season. On account of the energy and water-saving incentives to the farmers, shifting from flooded to merely moist soil and alternate wetting and drying system are observed to be the very attractive mitigation options. Moreover, biogas technology can also be used as excellent for controlling the emission of GHG and reducing global warming potential by providing biogas fuel in place of firewood and bio-fertilizer in place of chemical fertilizers. Currently, in India, around 3.83 million biogas plants are functional, which can overcome global warming by 38 Mt CO_2 eq. in a year and can generate the revenue through carbon credit under the clean development mechanism, further helping the farmers to develop biogas as a profitable venture (Pathak, 2015).

2.6.2 Mitigation of Nitrous Oxide Emission

The processes like nitrification and denitrification are responsible for the emission of nitrous oxide which can be reduced through the site-specific nutrient management practices and by the application of nitrification inhibitors (10-15%) like coated calcium carbide, neem cake, dicyandiamide, an extract of Karanja seed and neem oil. About 10-15%, emission can further be decreased by the use of some other techniques such as demand-driven N -using leaf colour chart and global warming potential (GWP) (Bhatia et al., 2012, Jain et al., 2013).

2.6.3 Carbon Sequestration in Agricultural soils

Sequestration of carbon is the seizure and storage of carbon in the water, soil, vegetation and the engineering structure. Carbon sequestration is based on the following main considerations: (i) whole soil profile (ii) long term storage of carbon (iii) regional-scale assessment (iv) assessment of life cycle of all the components of the ecosystem (v) inclusion of all the greenhouse gases and all form of carbons.

By effective management of soil, an appropriate amount of carbon can be sequestered. For which firstly, existing carbon is preserved in the soil by reducing the rate of organic matter decomposition with reduced-tillage and preventing soil erosion (Manyapu et al., 2019). Secondly, amount of carbon is increased by retaining crop residue, applying bio char as slow degrading carbon source, using perennial crops with a more extensive root system and balanced and integrated use of plant nutrient system (Lal, 2010, Smith et al., 2013, Paustian et al., 2016). Ultimately such approaches increase the sequestration of carbon in soil, which indicates the storage of carbon as soil organic matter (Lal, 2004, Pathak et al., 2011). Thus, stored soil organic carbon (SOC), mitigates the emission of CO_2 and also enhances the productivity of soil. An adequate supply of nutrients as organic manure and compost to soil can enhance SOC pool with biomass production higher than the use of the equal quantity of nutrients as inorganic fertilizers. However, SOC sequestration in tropics and the sub-tropics soil is a big challenge, where the climate is severe and agricultural inputs such as crop residues and organic manure cannot be afforded by the resource poor farmers. But, there is cheerfulness, that agricultural soil sequestration will at last be part of any long-term international convention for reductions in net GHGs emissions.

2.6.4 Conservation Agriculture

Resource-conserving technologies such as labour, water and nutrient saving techniques, permanent/semi-permanent residue cover, minimum tillage with direct seeding and crop rotations in conservation agriculture have an excellent potential for improving the sustainability of agriculture by the sustainable use of natural resources such as air, water, soil and fossil fuels (Chaplota et al., 2012, Gonzalez-Sanchez et al., 2012, Friedrich et al., 2014, Gao et al., 2016, Paliya et al., 2019). The adoption of such type of resource-conserving technologies in conservation agriculture has been proved very helpful in saving inputs, increasing profits, reducing wastes and in the mitigation of

GHGs emissions. The use of agricultural feedstock such as energy crops, crop residues and cow dung in energy production can reduce the greenhouse gases emission by substituting fossil fuels. To encourage the farmers for adoption of these mitigations techniques to take advantage of improved soil health and water and energy efficiency, some policies and incentives to be developed (Pathak et al., 2014).

2.6.5 Genetic Improvements of Crops and Animals

Tremendous improvement in production of food has been achieved by the adoption of genetic engineering, improved breeding techniques, appropriate genetic selection and the development of new varieties. Such type of techniques can improve the traits such as drought and pest resistance, product quality, nutritional benefits, flavour profile and sequestration of carbon, while reducing input requirements, water content, losses from waste and embedded carbon. Genetic engineering will produce essential perennials and tree crops like – bananas, oranges, palm oil, sugarcane, coffee, plantations, nuts *etc.* which helps in the mitigations of greenhouse gases by sequestering the carbon. Animal breeding and genetics are also used for increasing the available calories. Currently, in such techniques, most of the work is mainly focused on the reduction in anti-nutritional in feed, increment in palatability of feed and feed conservation techniques instead of changing the animals themselves. Moreover, to decrease the contribution of livestock to greenhouse gas emissions, ample work is being taken on to decrease the amount of CH_4 released by them.

2.6.6 Increasing Input-use Efficiency

In the agricultural production, the agricultural inputs such as labour, pesticides, nutrients and water are used as the significant factors for production. In Indian agriculture, the effectiveness of such inputs is low, which causes a large amount of losses and greenhouse gas emissions along with a decline in the production of some crops. By increasing the input use efficiency through appropriate resource-conserving technologies, emission of greenhouse gases can be mitigated in agricultural production, for example by increasing irrigation efficiency through modern most efficient resource-conserving irrigation technologies such as modified centre pivot, sprinkler and underground drip, 50% more water can be provided in agricultural production to encounter the overall food demand. In the case of nutrients application, especially nitrogenous fertilizers, the emission of nitrous oxide is direct. Still, the emission of carbon dioxide is indirect as in the manufacturing of this fertilizer natural gas is used as the main ingredient. Indirect emission of greenhouse gases can be reduced by lowering the consumption of nitrogenous fertilizer and by raising the emission intensity of fertilizer production, particularly of urea by replacing the old and less efficient technologies. But there are some difficulties in putting them in practice due to various rigidities in manufacturing, financing investments and policy change processes. Such type of difficulties shifts towards other mitigation options of reducing the use of nitrogenous fertilizer and increasing the efficiency of nitrogen-use. The use of cyanobacteria may influence nitrogen use, inter-cropping with legumes, leaf colour chart and deep embedded super granular urea. Cyanobacteria can fix the nitrogen in the soil and reduce nitrogenous fertilizer dependency from external sources. Similarly, the use of leaf colour chart become helpful in the understanding of the appropriate application of nitrogenous fertilizers. *i.e.*, it helps the farmer to understand when the application is less than optimal and when it is overdosed (Pathak et al., 2014 and 2015, Pathak, 2015).

2.7 Adaptation Option for a Climate Change and Their Co-benefits for Mitigation

Potential threats are related with climate change due to which a sincere effort on identifying, analyzing and understanding the adaptation is required. Societies and individuals are vulnerable to climate risks, and such vulnerability may act as a driver for adaptive resource management. Some adaptations are to be undertaken by the government on behalf of society and others by the individuals in response to climatic threats.

Mitigation and adaptation are generally looking over as different activities. Mitigation aims to decrease emission of greenhouse gases and adaptations help to adjust the expected increases in GHGs emission. Among these activities, the key concept is the identification of successful adaptations where the biggest risk and physical vulnerability persists. However, in agriculture, significant mitigation effects can also be generated by the adaptation measures for making them a greatly productive investment. It is crystal clear that collective efforts are needed for adaptation and mitigation to decrease the susceptibility of Indian agriculture to the destructive effect of climate change and making it more flexible. However, adaptation opportunities are immediately required, whereas mitigation options take a considerable time to affect climate change. Generally, the low adaptive capacity of poor farmers is observed due to their low formal education and subsistence agriculture. Because of this, a very simple, culturally acceptable and financially viable adaptation approaches have to be developed and implemented. Several high location-specific mitigation co-benefits of such adaptation approaches have been observed and are presented in Table-4.

Table 4: Adaptation opportunities for climate change in Indian farming with their mitigation co-benefits

Opportunities	Details	Adaptation advantages	Mitigation co-benefits
Crop diversification	Growing crops adjusting to the adverse climate	• Increased and stable production • Less water and energy use	• Reduced greenhouse gases due to proficient use of agri-inputs in the life cycle
Crop insurance	Incentives for the farmers to cover extremes of climates	• Increasing risk-taking ability • Livelihood security	-
Conservation agriculture	• Crop rotation • Zero tillage • Residue cover of soil	• Improved health of the soil • Less water and energy requirements	• Carbon sequestration • By efficient use of agricultural inputs GHGs emission can be reduced
Organic farming	Hindering the use of chemical based pesticides and fertilizers Promoting the use of an organic source of nutrients	Less energy requirements Improved soil health	• Carbon sequestration • Reduced CO_2 with less energy-use for pesticides and fertilizer manufacturing
Integrated pest management	Integrating the biological, chemical and physical pest management methods	• Increased crop yield with minimum loss by the pest infections	• Reduced carbon dioxide with less energy-use for pesticide manufacturing
Water-saving technologies	• Laser aided land levelling • Fertigation • Drip and sprinkler method	• Increased production and income • Water-saving • Enhanced nutrient use efficiency	• Reduced methane due to no submergence in rice cultivation • Decreased carbon dioxide due to less water use • Decreased nitrous oxide due to increased nitrogen use efficiency
Changing the date of planting	Avoiding heat stress during flowering and maturity through adjusting the planting date	• Less use of water and energy • Reduced yield loss	• Reduced carbon dioxide with less energy for irrigation
Climate-resilient crop varieties	Crop varieties heat, drought and flood of shorter duration of high yield	• Increased income • Saving energy and water • Stable and increased production	• Reduced greenhouse gases due to the shorter duration • Reduction in carbon dioxide due to less energy use for irrigation

Opportunities	Details	Adaptation advantages	Mitigation co-benefits
Creation of seed banks	• Arranging quality seeds to the farmers • Providing seeds in case of crop failure	• Enhanced farm income • Modified nutrient use efficiency	• Reduced greenhouse gas emission per unit production
Protected cultivation	• Cultivation of crops in the poly house for control of pest, moisture and temperature	• Increased stable production • Decreased climatic risk	-
Intercropping/mix cropping	cultivation of more than one crop on the same field for avoiding crop failure and for increasing crop productivity	• Improved nutrient use efficiency • Increased farm income	• Reduced nitrous oxide and carbon dioxide production
Wasteland management	Developing wastelands for grassland, forestry, agroforestry and crop production with improved nutrient and water management	• Improved soil health • Increased income • Increased crop production • Livelihood security	• Carbon sequestration
Upgraded weather-based agro-advisory	Information related to weather forecasting become very helpful in planning the crop management	• Reduced risk • Increased and stable production	-
Upgraded nutrient management	Site-specific, demand-driven and balanced use of nutrients	• Increased yield • Increased nutrient use efficiency	• Decreased nitrous oxide because of high nitrogen use efficiency.
Rainwater harvesting	• Collection and storage of rainwater • Utilization of harvesting water for irrigation • Recharging of groundwater	• Enhanced water availability • Reduced nutrient loss • Stable and increased yield	• High yield and efficient use of agri-inputs reduced GHGs emissions.
Use of effective microbes	Improving soil fertility with crop productivity	• Improved soil health and increased crop yields	• Carbon sequestration • Decreased emission of greenhouse gases (nitrous oxide and carbon dioxide)
Precision farming	Precise and site-specific management of agri-inputs	• Enhanced yield • Enhanced input use efficiency	• Decreased greenhouse gases emission due to efficient use of agri-inputs

Opportunities	Details	Adaptation advantages	Mitigation co-benefits
Use of native technical knowledge	Applying farmers indigenous technical knowledge in weather forecasting and risk management in crop cultivation	• Stable and increased production • Reduced risk	-
Use of bio-fuels	Use of biofuels from non-edible crops and crop residue in conjunction with fossil fuels	• Increased farm income • Decreased dependence on fossil fuel	• Carbon sequestration with the growing of perennial non-edible crops
Use of non-conventional energy	Substitution of fossil fuel-based conventional energy sources by wind and solar energy	• Increased farm income • Reduced dependence on imports	• Decreased carbon dioxide with lesser use of fossil fuels
Agro-horticulture, agroforestry	Crop production along with the horticulture and forestry	• Increased income • Livelihood security	• Carbon sequestration
Cooperative farming	Farmers joining together for cultivation and marketing of produce farming, to acquire latest approaches and bear greater risks compared to one farmer	• Increased income • Stable and increased crop production • Increased efficiency of resources	• Decreased greenhouse gases with increased efficiency of agri-inputs
Intensifying crop production	Improved crop production by intensive use of irrigation and fertilizers	• Greater efficiency of inputs • Stable and increased production	• Reduced greenhouse gases *per* unit production

2.8 Economic Implications of GHGs Mitigation

Twenty different rice production technologies were analyzed for their GWP and associated cost (Pathak and Wassmann, 2007). Such technologies represented separate options for mitigation of GHG emissions due to their different water regime, soil and residue management, residue utilization and additives practices. Thirteen out of 20 technologies were observed to have the ability to decrease the GWP by 8% to 51%, in comparison to the farmers' present practice of rice cultivation. The replacement of application of urea plus farmyard manure (FYM) and continuous flooding by the application of urea alone and alternate flooding will reduce the global warming potential (GWP) by 29% and 15%, respectively. However, in comparison to the present practice of rice crop residue burning and using farmyard manure, supplying of nitrogen through urea and feeding of rice straw to cattle will reduce it by 41%. Amendment of inorganic nitrogen fertilization with phospho-gypsum and nitrification inhibitor was noticed as the most promising mitigation option resulting in shadow rates of < 10 US$ *per* ton of CO_2 eq. (Wassmann and Pathak, 2007). As compared to alternate flooding, mid-season drainage had a better cost-benefit however, was observed less profitable than rice husk utilization. But in most of the cases, biogas technology was observed as the preferable option. In another study, 16 technologies have been assessed (Pathak, 2010), which are recommended for increasing N-use efficiency, for their GHGs mitigation potential and economic returns. These technologies reduced GWP by 1-9%. GWP can be reduced without the involvement of any extra cost by the use of resource-conserving technologies, other flooding technologies and mid-season drainage approach.

The approach, such as integrated nutrient management, increases the GWP. The technologies such as intermittent drying and site-specific N management can quickly be adopted by the farmers as no additional investment is required in these techniques. In contrast, other technologies required monetary incentives and policy support. The greenhouse gases mitigation ability and constraints of some most promising techniques are presented in Table 5.

Table 5: Mitigation ability and constraints of greenhouse gas mitigation techniques

Sr.No	Source	Mitigation techniques	Mitigation ability (%)	Constraints
1	Carbon sequestration in soil	• Conservation agriculture	15-20	Small holding, continuity
		• Organic farming	15-20	Cost, manure availability
2	Methane from ruminants	• Efficient animals	10-20	Cost, acclimatization
		• Feed additives	5-10	Cost, bio safety
		• Balance feeding	5-10	Open grazing, cost
3	Nitrous oxide from soil	• Nitrification inhibitor	10-15	Incentive cost
		• Site-specific N use	10-15	Awareness, fertilizer policy
4	Methane from rice field	• Irregular drying	25-30	Secure irrigation
		• SRI	20-25	Secure irrigation, labour
		• Direct seeded rice	30-40	Herbicides and machines

Sources: Lungerska and Chakir, 2018, Pathak et al., 2010, Talayer, 2018

2.9 Trade-offs between GHGs Emission and Crop Production

In India, the requirement for rice and wheat is expected around 77 % of total food grains production and also look forward to increase in the coming future. By using the increased inputs, especially irrigation and fertilizers, this increased demand may be fulfilled. However, the use of increased inputs may contribute to the increased emission of greenhouse gases. Therefore, to enhance food production and to reduce GWP by increasing the carbon efficiency ratio and by developing suitable technologies, there is a requirement for setting a trade-offs between crop production (wheat and rice) and GWP. Supply and demand driven irrigation with ten levels of nitrogen and organic manure and Info Crop model have been used (Bhatia et al., 2010) to simulate the yield and GWP of soils in wheat and rice crop for five regions of the country for two climatic situations of 1990-1999 and 2050. Such type of simulation study has estimated the yield response and GWP of crops (rice and wheat) by the high nitrogen and water use in present and future climatic situations. The water and nitrogen availability limited the current yield of rice and wheat. Increased irrigation and nitrogen use would increase the productivity, carbon efficiency ratio and GWP of rice and wheat from the current value of 3.26 Mg ha-1 and 2.73 Mg ha-1 to 5.66 Mg ha-1 and 6.15 Mg ha-1; 0.67 and 0.85 to 1.06 and 1.75 and by 27% and 40%, respectively. Therefore, more reasonable management techniques are required for the GHGs emissions mitigation and an increase in crop produce and carbon efficiency ratio by the efficient use of inputs, especially nitrogen and irrigation.

2.10. Options for Carbon Trading in Indian Agriculture

The targets to reduce the emissions of greenhouse gases, committed by the countries under the Kyoto Protocol must meet primarily through the national measures. Additionally, for meeting these targets, three market-based Kyoto mechanisms were announced which includes Clean Development Mechanism (CDM), Joint Implementation (JI), and International Emission Trading (IET). Such types of mechanisms are now known as “carbon trading”. The Joint

Implementation is a project-based mechanism that feeds the carbon market and permits the industrial countries to execute projects with other developed countries as joint implementation projects. Such mechanisms permit the developed countries to trade GHG emission reduction units get from their mitigation units to other developed countries. Clean Development Mechanism is a project-based mechanism which permits the industrial countries to finance sustainable management projects which decrease emissions in developing countries and receive credit for this. Trading of emissions-reduction credits involve the International Emission Trading mechanism. A country which has less entitlement to produce than its real emissions could acquire credits to overcome its shortfall. To increase the sustainability of the agricultural production, farming community carried out several agricultural activities which may potentially qualify the Kyoto mechanisms and gain the carbon credits. Such type of trading increases the farmers’ income which may invite investments benefiting sustainable development and the population at large (Pathak, 2015).

2.11. Research and Policy Opportunities for GHGs Mitigation in Indian Agriculture

The appropriate adaptation and mitigation strategies have to be developed for making the agriculture sustainable and climate-resilient and for ensuring the short- and long-term food security of the country. India has launched a capacity building program on research and development in climate-resilient agriculture during February 2011 named as “National Initiative on Climate resilient agriculture (NICRA)” for the enhancement of agricultural productivity and the reduction in an adverse impact on the environment. The main objective of this program is to establish the required infrastructure to carry out the research and to train the agricultural scientists on climate-change-resilient related research, crops, fisheries and livestock. National Mission for Sustainable Agriculture (NMSA) has also been started by India during April 2014 which involves the issues regarding sustainable agriculture, short- and long-term food security, climate-resilient, appropriate adaptation and mitigation strategies *etc.* Sustainable agricultural production is proved as a key to food and livelihood security of the country. Agriculture provides effective greenhouse gases mitigation opportunities through biomass production, land use management and carbon sequestration (Pathak et al., 2014). So, some policy including avenues such as investments, incentives and information should be framed to encourage farmers to adopt the mitigation approaches without compromising the income and production. In Indian agriculture, to promote the GHG mitigations, some research and policy options are designed as:

a. Issues for further research

i. Investigating the protocols for measurement, reporting and verification (MRV) system for emission and mitigation of GHGs.

ii. Identifying the mitigation co-benefits of climate change adaptation technologies.

iii. Generating simple techniques for quantifying the emission of GHGs from agriculture and decreasing uncertainties in emission coefficients.

iv. Evaluating the emission of GHGs from large agricultural areas in the life cycle of agricultural produce.

v. Elucidating the technical and economic feasibilities of different greenhouse gas mitigation approaches.

vi. Generating low nitrogen and carbon technologies and investigating their potential for greenhouse gases mitigation.

vii. Formulating simulation models for integrated determination of greenhouse gases emission and coupling mitigation with remote sensing and geographical information system (GIS).

viii. Generating techniques for reducing GHGs emissions from livestock by better manure management and feeding techniques.

b. Policy issues

i. Generation of awareness for the reduction of GHGs in agriculture by adopting best management practices.

ii. Investing in rural infrastructure to improve the resilience of agriculture in the face of uncertainties of climate change.

iii. Developing the GHGs inventory at the regional and national level by making the institutional mechanism for the gathering and management of data.

iv. Financing the research stringing the understanding of interactions between climate change and agriculture.

v. Framing the innovative payment system and support for mitigation for small landholder farmers.

vi. Assisting the farmers regarding new carbon markets, new regulatory structures, government priorities and policies.

vii. Introducing the concept of green budgeting at the state and national levels by linking subsidy on agri-inputs with the mitigation of greenhouse gases.

2.12 Conclusions

There is no doubt that agriculture is one of the prime sources of greenhouse gas emission. Still, it has ability to mitigate the GHGs emission by carbon sequestration as well as by reduction in the emission of methane and nitrous oxide. Despite significant mitigation potential of several agricultural practices, the inclusion of agriculture in the emissions offset markets and climate change policy is not done due to inadequate information and lack of confidence in observing of agricultural interventions. By complying the agricultural mitigation projects with the established methodologies, additional reductions in GHGs emissions or carbon sequestration may be provided via the active participation in emissions offset markets. For quantifying agricultural mitigation practices, small work has been conducted towards developing sound technologies. A robust understanding of extent of greenhouse gases can be reduced or extent of carbon can be sequestered by different mitigation practices focuses on the decision on the most appropriate mitigation strategies. The effective agricultural GHG mitigation will require additional resources, advanced and cost-effective financing, multiple funding streams, unequivocal eligibility of agriculture, including sequestration of soil carbon in prevailing financing system. The effective agricultural GHG mitigation will also require enhanced experimentation on capacity building, mitigation and adaptation and modifications in land use management. The pool of approved methodologies seeks the financial rewards to the farmers for acceptance of climate-friendly practices. With the emerging profile of agriculture in the UNFCCC process, there is a need to invest in the improvement of agricultural MRV systems that are useful for the varying public and private approaches for providing with an incentive for agricultural mitigation. Ultimately to make the Indian agricultural production systems viable, there is a requirement for searching appropriate soil management system in which organic carbon of the soil is enhanced, and greenhouses gases emission is reduced. To encourage the farmers for the sustainable use of resources (water and energy) and to adopt improved mitigation options for the improvement of soil health, policies and incentives should be developed.

ORCID

Asha Gupta http://orcid.org/0000-0002-2512-8305

Hardeep Rai Sharma http://orcid.org/0000-0002-5130-5205

References

Abbas, A., Hammad, H. M., Ishaq, W., Farooque, A. A., Bakhat, F., Zia, Z., Fahad, S., Farhad, W., Cerda, A., 2020. A review on soil carbon dynamics resulting from agricultural practices. Environmental Management. 268, 1–16.

Albrecht, A., Kandji, S. T., 2003. Carbon sequestration in tropical agroforestry systems. Agricultural Ecosystems and Environment. 99, 15–27.

Bharti, V., Vikrant, K., Goswami, M., Tiwari, H., Sonwani, R.K., Lee, J., Tsang, D.C.W., Kim, K.H., Saeed, M., Kumar, S., Rai, B.N., Giri, B.S., Singh, R.S., 2019. Biodegradation of methylene blue dye in a batch and continuous mode using biochar as packing media. Environ. Res. 171, 356–364. https://doi.org/10.1016/j.envres.2019.01.051

Bhatia, A., Aggarwal, P.K., Jain, N., Pathak, H., 2012. Greenhouse gas emission from rice and wheat-growing areas in India: Spatial analysis and up scaling. Greenhouse Gases Science and Technology. 2, 115–125.

Bhatia, A., Jain, N., Pathak, H., 2013. Methane and nitrous oxide emissions from Indian rice paddies, agricultural soils and crop residue burning. Science and Technology. 3, 196–211.

Bhatia, A., Pathak, H., Aggarwal, P. K., Jain, N., 2010. Trade-off between productivity enhancement and global warming potential of rice and wheat in India. Nutrient Cycling in Agro ecosystems. 86, 413–424

Biswas, A.K., Kumar, S., Babu, S.S., Bhattacharyya, J.K., Chakrabarti, T., 2010. Studies on environmental quality in and around municipal solid waste dumpsite. Resour. Conserv. Recycl. 55, 129–134. https://doi.org/10.1016/j.resconrec.2010.08.003

Bouwman, A., 2001. Global estimates of gaseous emissions from agricultural land. Rome, Italy: FAO.

Cannell, M. G. R., 2003. Carbon sequestration and biomass energy offset: theoretical, potential and achievable capacities globally, in Europe and the UK. Biomass Bioenergy. 24, 97–116.

Cassman, K. G., Dobermann, A., Walters, D. T., Yang, H., 2003. Meeting cereal demand while protecting natural resources and improving environmental quality. Annual Review of Environmental Resources. 28, 315–358.

Chaplota ,V., Mchunub, C.N., Mansonc, A., 2012. Water erosion-induced CO_2 emissions from tilled and no-tilled soils and Sediments. Agriculture Ecosystems and Environment. 159, 62–69.

Clemens, J., Ahlgrimm, H. J., 2001. Greenhouse gases from animal husbandry: mitigation options. Nutrient Cycling Agroecosystems. 60, 287–300.

De-Pinto, A., Cnacchi, N., Kwon, H. Y., Koo, J., Duston, S., 2020. Climate smart agriculture and global food crop production. PLoS ONE. 15(4), 1–19.

Dhar, H., Kumar, S., Kumar, R., 2017. A review on organic waste to energy systems in India. Bioresour. Technol. 245, 1229–1237. https://doi.org/10.1016/j.biortech.2017.08.159

Eidman, V. R., 2005. Agriculture as a producer of energy. In Agriculture as a producer and consumer of energy (Outlaw, J. L., Collins, K. J. and Duffiield, J.A. eds.), Wallingford, UK: CAB International. 30–67.

Fagodia, R. K., Pathak, H., Bhatia, A., Jain, N., Kumar, A., Malyan, S. K., 2020. Global warming impact of nitrogen use in agriculture: An assessment of India since 1960. Carbon Management. 11(3), 291–301.

FAI: The Fertilizers Association of India, 2017. Available at: https://www.faidelhi.org/.

FAO (Food and Agriculture Organization of United Nations), 2018. The State of Food Security and Nutrition in the World 2018. Building climate resilience for food security and nutrition. Rome, FAO.

FAO, 2017. Climate smart agriculture sourcebook. Available at: https://fao.org/climate-smart-agriculturesourceebook/en/.

Feliciano, D., Nayak, D. R., Vetter, S.H., Hiller, J., 2017. CCAFS-MOT-a tool for farmers, extension services and policy advisors to identify mitigation options for agriculture. Agricultural System. 154, 100–111.

Foley, J. A., De Fries, R., Asner, G.P., Barford, C., Bonan, G., Carpenter, S.R., Chapin, F., Coe, M.A., et al., 2005. Global consequences of land use. Science. 309. 570–574.

Frank, S., Havlik, P., Soussana J.F., 2017. Reducing greenhouse gas emissions in agriculture without compromising food security? Environmental Research Letters.12, 105004.

Friedrich, T., Kassam, A., Corsi, S., 2014. Conservation agriculture: Global prospects and challenges: conservation agriculture in Europe. CABI.

Galloway, J. N., Aber, J. D., Erisman, J. W., Seitzinger, S. P., Howarth, R. W., Cowling, E. B., Cosby, B. J., 2003. The nitrogen cascade. Bioscience. 53, 341–356.

Gao, Y., Dang, X.H., Yu, Y.,Li, Y., Liu, Y., Wang, J., 2016. Effects of tillage methods on soil carbon and wind erosion. Land Degradation Development. 27, 583–591.

Ghosh, A., Mishra, S., Bhattacharya, R., Sarkar, A., Singh, A. K., Tyagi, V. C., Kumar, R. V., Meena, V. S., 2020. Agriculture, dairy, fishery, farming practices and greenhouse gas emission foot print: a strategic appraisal for mitigations. Environmental Science and Pollution Research. 27, 10160–10184.

Gibbs, H. K., Ruesch, A. S., Achard, F., Clayton, M. K., Holmgren, P., Ramankutty, N., Foley, J. A., 2010. Tropical forests were the primary sources of new agricultural land in the 1980s and 1990s. Proceedings of the National Academy of Sciences. 107, 16732–16737.

Gonzalez-Sanchez, E.J., Ordonez-Fernandez, R., Carbonell-Bojollo, R., Veroz-Gonzalez, O., Gil-Ribes, J. A., 2012. Meta-analysis on atmospheric carbon capture in Spain through the use of conservation agriculture. Soil Tillage. 122, 52–60.

Gupta, P.K., Gupta ,V., Sharma, C., Das, S.N., Purkait, N., Adhya, T.K., Pathak, H., Ramesh, R., Baruah, K.K., Venkatratnam, L., Singh, G., Iyer, C.S.P., 2009. Development of methane emission factors for Indian paddy fields and estimation of national methane budget. Chemosphere. 74(4), 590–598.

Harvie, J., Schettler, T., Mikkelsen, L., Flora C., 2011. Common drivers - Common solutions: Chronic disease, Climate change, Nutrition and agriculture. Food systems and public health conference work team funded by the W.K. Kellogg Foundation, 49.

Hazell, P., Wood, S., 2008. Drivers of change in global agriculture. Philosophical.Transactions. of the Royal.Society B. 363, 495–515.

ICAR-CSSRI, 2016. Reclamation of alkali soil through gypsum technology folder. Central Soil Salinity Research Institute (CSSRI), Karnal, India.

IEA (International Energy Agency), 2018. World Energy Outlook: Analysis and key findings. A report by the International Energy Agency.

International Panel on Climate Change (IPCC), 2001. Climate Change 2001: The scientific basis. Contribution of working group I to the third assessment report of the intergovernmental panel on climate change. Cambridge, UK: Cambridge University Press.

IPCC, 2014. Climate Change: Mitigation of climate change. Contribution of Working Group III to the Fifth assessment report of the intergovernmental panel on climate change (Edenhofer, O., Pichs-Madruga, R., Sokona, Y., Farahani, E., Kadner, S., Seyboth, K., Adler, A., Baum, I., Brunner, S., Eickemeier, P., Kriemann, B., Savolainen, J., Schlömer, S., Von Stechow,

C., Zwickel, T., Minx, J.C. eds) Cambridge University Press, Cambridge, United Kingdom and New York, NY, USA.

IPCC, 2014. Impacts, adaptation and vulnerability working group-II Contribution to the fifth assessment report. Cambridge, UK: Cambridge University Press.

Jain, N., Bhatia, A., Pathak, H., 2014. Emission of air pollutants from crop residue burning in India. Aerosol and Air Quality Research. 14, 422–430.

Jain, N., Dubey, R., Dubey, D. S., Singh, J., Khanna, M., Pathak, H., Bhatia, A., 2013: Mitigation of greenhouse gas emission with system of rice intensification in the Indo-Gangetic Plains. Paddy and Water Environment. 12, 355–363 (2014)

Jaiswal, B., Agarwal, M., 2019. Carbon footprint of Agricultural sector. Environmental Footprints and Eco-Design of Food Products. 81–99.

Janzen, H. H., 2004. Carbon cycling in earth systems a soil science perspective. Agricultural Ecosystems & Environments. 104, 399–417.

Kumar, S., Negi, S., Mandpe, A., Singh, R.V., Hussain, A., 2018. Rapid composting techniques in Indian context and utilization of black soldier fly for enhanced decomposition of biodegradable wastes - A comprehensive review. J. Environ. Manage. 227, 189–199. doi:10.1016/j.jenvman.2018.08.096

Lal, R., 2004. Soil carbon sequestration impacts on global climate change and food security. Science. 304, 1623–1627.

Lal, R., 2004. Soil carbon sequestration in India. Climatic Change. 65, 277–296.

Lal, R., 2010. Managing soils and ecosystems for mitigating anthropogenic carbon emissions and advancing global food security. Bio Science. 60, 708–721.

Linquist, B. A., Adviento-Borbe, M. A., Pittelkow, C. M., van Kessel, C., van Groenigen, K. J., 2012. Fertilizer management practices and greenhouse gas emissions from rice systems: A quantitative review and analysis. Field Crops Research. 135, 10–21.

Liu, D., Guo, X., Xiao, B., 2019. What causes growth of global greenhouse emission? Evidence from 40 countries. The Science of the Total Environment. 661, 750–766.

Lungerska, A., Chakir, R., 2018. Climate induced land change in France: Impacts of agriculture adaptations and climate change mitigation. Ecological Economics. 147, 134–154.

Manyapu, V., Mandpe, A., Kumar, S., 2019. Corrigendum to "Synergistic effect of fly ash in in-vessel composting of biomass and kitchen waste" [Bioresour. Technol. 251 (2018) 114–120](S0960852417321740)(10.1016/j.biortech.2017.12.039). Bioresour. Technol. 277, 225. doi:10.1016/j.biortech.2018.12.031

Miller, N., Urrea, A., Kahmark, K., Shcherbk, I., Robertson, G.P., Oritz-Monasterio, I., 2018. Nitrous oxide flux response exponentially to nitrogen fertilizers in irrigated wheat in Yaqui valley, Mexico. Agricultural Ecosystems and Environment. 261, 125–132.

MoEFCC India, 2018. Second biennial update report to the United Nations Framework Convention on Climate Change. New Delhi: Ministry of Environment Forest and Climate Change, Government of India. 240.

Mosier, A. R., Duxbury, J. M., Freney, J. R., Heinemeyer, O., Minami, K., Johnson, D. E., 1998. Mitigating agricultural emissions of methane. Climate Change. 40, 39–80.

Oenema, O., Wrage, N., Velthof, G. L., van Groenigen, J. W., Dolfiing, J., Kuikman, P. J., 2005. Trends in global nitrous oxide emissions from animal production systems. Nutrient Cycling Agroecosystems. 72, 51–65.

Oertel, C., Matschullat, J., Zurba, K., Zimmermann, F., Erasmi, S., 2016. Greenhouse gas emission from soil- A review. Cheme der Erde. 76, 327–352.

Olivier, J.G.J., Schure, K.M. , Peters, J. A. H. W., 2017. Trends in global CO_2 and total greenhouse gas emissions: 2017 report. PBL Netherlands Environmental Assessment Agency, The Hague.

Paliya, S., Mandpe, A., Kumar, S., Kumar, M.S., 2019. Enhanced nodulation and higher germination using sludge ash as a carrier for biofertilizer production. J. Environ. Manage. 250, 109523. doi:10.1016/j.jenvman.2019.109523

Pathak, H., 2010. Mitigating greenhouse gas and nitrogen loss with improved fertilizer management in rice: quantification and economic assessment. Nutrient Cycling in Agro ecosystems. 87, 443–454.

Pathak, H., 2015. Greenhouse gas emission from Indian agriculture: Trends, Drivers and Mitigation strategies. Proceeding of Indian Science Academy. 81, 1133–1149.

Pathak, H., Adhya, T. K., Sikka, A. K., Islam, A., 2015. Water and climate change. In: State of Indian Agriculture: Water (Pathak, H., Bhatt, B. P. and Gupta, S. K., eds) National Academy of Agricultural Sciences, New Delhi. xx + 103.

Pathak, H., Bhatia, A., Jain, N., 2014. Greenhouse Gas Emission from Indian Agriculture: Trends, Mitigation and Policy Needs. Indian Agricultural Research Institute, New Delhi, 1–39.

Pathak, H., Bhatia, A., Jain, N., Aggarwal, P. K., 2010. Greenhouse gas emission and mitigation in Indian Agriculture–A Review, In: ING Bulletins on Regional Assessment of Reactive Nitrogen, Bulletin No. 19 (ed. Singh, B.), SCON-ING, New Delhi, 1–34.

Pathak, H., Byjesh, K., Chakrabarti, B., Aggarwal, P. K., 2011. Potential and cost of carbon sequestration in Indian agriculture: Estimates from long-term field experiments. Field Crops Research. 120, 102–111.

Pathak, H., Saharawat, Y.S., Gathala, M., Mohanty, S., Ladha, J.K., 2009. Simulating environmental impact of resource-conserving technologies in the rice-wheat system of the Indo-Gangetic Plains. In: Ladha, J.K., Singh, Y., Erenstein, O., Hardy, B., (eds). Integrated crop and resource management in the rice-wheat system of South Asia. Los Banos (Philippines): International Rice Research Institute. 321–334.

Pathak, H., Sankhyan, S., Dubey, D.S., Bhatia, A., Jain, N., 2013. Dry direct-seeding of rice for mitigating greenhouse gas emissions: Field experimentation and simulations. Paddy and Water Environment. 11, 593–601.

Pathak, H., Wassmann, R., 2007. Introducing greenhouse gas mitigation as a development objective in rice-based agriculture: I. Generation of Technical Coefficients. Agricultural Systems. 94, 807–825.

Paustian, K., Cole, C. V., Sauerbeck, D., Sampson, N., 1998. CO_2 mitigation by agriculture: an overview. Climate Change. 40, 135–162.

Paustian, K., Lehmann, J., Ogle, S., Reay, D., Robertson, G. P., Smith, P., 2016. Climate-smart soils. Nature. 532, 49–57.

Phillips, L. R., Tanaka, L. D., Archer, W. D., Hanson, D. J., 2009. Fertilizer application timing influences greenhouse gas fluxes over a growing season. Journal of Environmental Quality. 38, 1569–1579.

Prather, M., Derwent, R., Ehhalt, D., Fraser, P., Sanhueza, E., Zhou, X. 1995. Other trace gases and atmospheric chemistry. In: Climate Change 1994: Radiative Forcing of Climate Change and an Evaluation of the IPCC IS 92 Emission Scenarios (eds Houghton, J.T., Meira Filho, L.G., Bruce, J., Lee, H., Callender, B.A., Haites, E et. al.) Cambridge University Press, Cambridge 73–126.

Ramaswamy, V., Boucher, O., Haigh, J., Haughustaine, D., Haywood, J., Myhre, G., Nakajima, T., Shi, G.Y., Solomon, S., 2001. Radiative forcing of climate change. In: Climate Change 2001: The Scientific Basis (eds Houghton, J.T., Ding, Y., Griggs, D.J., Noguer, M., Van der Linden, P.J., Da, X. et. al). Cambridge University Press, Cambridge 349–416.

Ren, F., Zhang, X., Liu, J., Sun, N., Wu, L., Li, Z., Xu, M., 2017. A synthetic analysis of greenhouse gases emissions from manure amended agricultural soil in China. Scientific Report. 7, Article No. 8123.

Sapkota, T. B., Jat, R. K., Singh, R. S., Jat, M. L., Bijarniya, D., Kumar. M., Singh, Y., Saharawat Y. S., Gupta, R. K., 2017. Soil organic carbon changes after seven years conservation of agriculture based rice-wheat cropping system in eastern Indo-Gangetic plains of India. Soil Use and Management. 33, 81-89.

Sapkota, T. B., Vetter, S. H., Jat, M. L., Sirohi, S., Shirsath, P. B., Singh, R., Jat, S. H., Smith, P., Hiller, J., Stirling, C. M., 2019. Cost effective opportunities for climate change mitigation in Indian agriculture. The Science of the Total Environment. 655, 1342–1354.

Schleussner, C. F., 2016. Differential climate impacts for policy - relevant limits to global warming: the case of 1.50C and 20C. Earth System Dynamics.7, 327–351.

Schneider, U. A., McCarl, B. A., 2003. Economic potential of biomass based fuels for greenhouse gas emission mitigation. Environmental and Resource Economics. 24, 291–312.

Sejian,V., Bhatta, R., Malik, P. K., Madiajagan, B., Al-Honsi, Y. A. S., Sullivian, M., Gaughan, J. B., 2016. Livestock as source of greenhouse gases and its significance to climate change. Greenhouse Gases, Bernardo Llamas Moya and Juan Pous, IntechOpen, DOI: https://doi.org/10.5772/62135

Sharma, B., Furooq, M., Gogoi, N., Borkotoki, B., Kataki, R., Gargi, A., 2018. Soil organic carbon dynamics in wheat-green gram crop rotation amended with vermicompost and biochar in combination with inorganic fertilizers: a comparative study. Journal of Cleaner Production. 201, 471–480.

Smith, K. A., Conen, F., 2004. Impacts of land management on fluxes of trace greenhouse gases. Soil Use Management. 20, 255–263.

Smith, P., 2004. Engineered biological sinks on land. In the global carbon cycle. Integrating humans, climate and the natural world (eds Field, C. B., Raupach, M. R.). Washington, DC: Island Press, pp. 479–491.

Smith, P., 2016. Soil carbon sequestration and biochar as negative emission technologies. Global Change Biology. 22, 1315–1324.

Smith, P., Haberal, H., Popp, A., Erb, K.H., Lauk, C., Harper, R., Tubiello, E.N. et al., 2013. How much land based greenhouse gas mitigation can be achieved without compromising food security and environmental goals? Global Change Biology. 19, 2285–2302.

SNS, 2012. Second National Communication to the United Nations Framework Convention on Climate Change, Ministry of Environment and Forest, Government of Haryana.

Talayer, M., 2018. Climate smart agriculture: What is it good for? Journal of Peasant Studies. 45(1), 89–107.

Tongwane, M., Mdlambuzi, T., Moeletsi, M., Tsubo, M., Mliswa, V., Grootboom, L., 2016. Greenhouse gas emission from different crop production and Management practices in South Africa. Environmental Development. 19, 23–35.

Tubiello, F. N. Salvatore, M., Ferrara, A.F., House, J., Federici, S. et al. 2015. The contribution of agriculture forestry and other land use activities to global warming, 1990–2012. Global Change Biology. 21(2), 655–660.

UNFCCC Decision of COP 23, 2018. Report of the Conference of Parties on its 23 session held in Bonn from 6-18 November 2017 Bonn. United Nation Framework Convention on Climate Change (UNFCCC).

UNFCCC Paris Agreement. United Nations Treaty Collections, 2015. available at https://treaties.un.org.

UNFCCC, 2008. Challenges and opportunities for mitigation in the agricultural soils. United Nations Framework Convention on Climate Change, Document No. FCCC/TP/2008/8.

Wainaina, S., Awasthi, M.K., Sarsaiya, S., Chen, H., Singh, E., Kumar, A., Ravindran, B., Awasthi, S.K., Liu, T., Duan, Y., Kumar, S., Zhang, Z., Taherzadeh, M.J., 2020. Resource recovery and circular economy from organic solid waste using aerobic and anaerobic digestion technologies. Bioresour. Technol. 301, 122778. https://doi.org/10.1016/j.biortech.2020.122778

Wang, J., Xi, F., Liu, Z., Bing, L., Alsaedi, A., Hayant, T., Ahamd, B., Guen, S., 2018. The spatiotemporal features of greenhouse gases emissions from biomass burning in China from 2000–2012. Journal of Cleaner Production. 8, 801–808.

Wassmann, R., Pathak, H., 2007. Introducing greenhouse gas mitigation as a development objective in rice-based agriculture: II. Cost-benefit assessment for different technologies, regions and scales. Agricultural Systems. 94, 826–840.

World Resource Institute Climate Analysis Indicators Tools (WRI CAIT 4.0)., 2017. Greenhouse gas emission is expressed in units of carbon dioxide equivalents. Global warming potentials (GWP) are the 100 years GWP from the Intergovernmental panel on Climate Change (IPCC) Second Assessment Report (SAR).

Yan, X., Ohara, T., Akimoto, H., 2003. Development of region-specifiic emission factors and estimation of methane emission from rice field in East, Southeast and South Asian countries. Global Change Biology. 9, 237–254.

Zhao, X., He, J., Cao, J., 2011. Study on mitigation strategies of methane emission from rice paddies in the implementation of ecological agriculture. Energy Procedia. 5, 2474–2480.

3

Waste Recycling

Mehak Puri[a,b], Sonam Paliya[ab*]

[a]Academy of Scientific and Innovative Research (AcSIR)
Ghaziabad- 201 002, India
[b]CSIR- National Environmental Engineering Research Institute
(CSIR-NEERI), Nehru Marg, Nagpur-440020, India

ABSTRACT

Waste is heterogenous in nature. Composition and quantification for the waste are very essential parameters. There are methods to acquire the waste sample prior the analysis. The analysis of wastes suggests number of options to reduce and get rid of the unwanted materials; collect and reuse the materials of importance. Most environment friendly and economical options turn out to be recycling. Recycling is widely practiced all over the world. The approach available nowadays is to recycle every item as much as possible to reduce burden on the environment and even considering the economic perspective. The chapter deals with different methods of recycling and the economic analysis touching the tangibles and intangibles. Very few wastes are discarded and sent to landfill after recycling. Majority is reverted back and reused as a substitute of virgin raw materials to make new products. The highlight of the chapter is recycling is one of the most inexpensive, easy and convenient methods of all. Increase of public participation has led to reduction of the waste and it is reused till the end of life of the material.

Keywords: Waste management, Waste recovery, Waste characterization, Recycling, Cost Analysis, Life Cycle Assessment

3.1 Introduction

A Massive increase in population, urbanization and industrialization have led to an increase in the quantity and composition of waste. The waste is increasing day by day and is estimated to reach up to 2.2 billion tonnes by 2025, according to the reports by World Bank. It becomes a necessity for every nation to fore look upon the waste management system for their respective countries. From the collection, segregation, transportation, disposal all have become an important criterion to deal with. Among the available disposal options; the easiest and environment friendly is "Recycling". Recycling is the process of reusing the materials in a way so as to substitute the use and extraction of the virgin or fresh materials that are required to make new products. The idea is to gain maximum output from the waste in the form of energy or fuel production. Reduce the burden on the landfill and incinerators, which removes the impact on the environment leading to a "sustainable approach". For attaining maximum output from the waste, it is imperative to analyze the composition in terms of quantity and quality. It helps us characterize the waste, so it becomes simpler to assess the waste fit for recycling. Glass, metals, plastics, cardboards, papers, *etc.* are very common materials that are recycled. E-waste has gained attention. The mix of metals, plastics, chemicals is all composition of e-waste. Hence some materials are recyclable and some are considered as hazardous waste. The situations like these reflect the importance to characterize the waste.

As mentioned above the use of recycling. This is just not limited to the environmental aspect, but even the economic aspect is considered. The chapter deals with analyzing step-wise cost estimation of the recycling process. The importance of public participation is evident in the success of the recycling initiatives.

3.2 Waste Recovery

Waste which is unwanted, unpleasant, full of odour expected with short life span and hence discarded as rubbish, garbage, *etc.* into the environment. An increase in population leads to an increase in demand for resources and goods. More the consumption more will be the waste generation. Presently about 2.01 billion tonnes annually of waste are generated, which is expected to rise as high as 3.40 billion tonnes by 2050 *i.e.,* scaffold increase of 70%. The global footprint in waste generation is 0.74 kilograms per person per day. Presently the world is able to recycle 13.5% of its waste and 5.5% is used as compost (world bank report "What a Waste 2.0" **(Hoornweg and Bhada-Tata 2012)**. Waste Recovery is one of the least preferred waste management methods ***(fig 1*** that utilizes the materials from the waste (*i.e.,* virgin resources) to generate valuable end product resources (heat, electricity, compost and fuel).

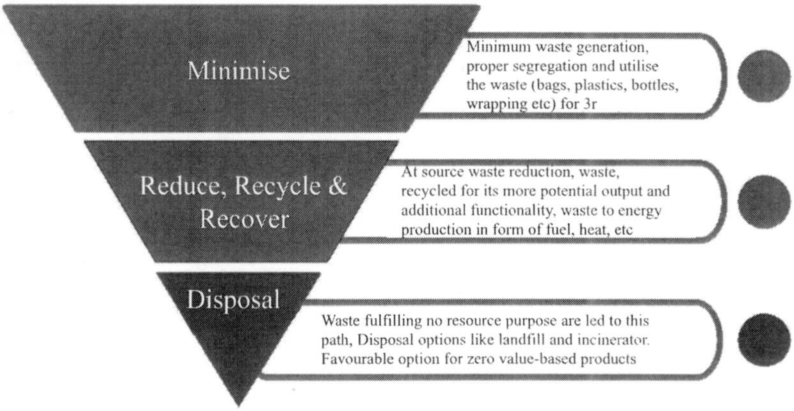

Fig. 1: Waste management hierarchy (Source: Ministry of Housing and Urban Affairs, 2017)

Fig.1. It illustrates the waste management trend followed for every waste. The waste recovery is one less to the last disposal option *i.e.,* maximum potential utilisation of the waste is carried out. It helps to mitigate the release of greenhouse gases (GHG's) and minimise the load of waste reaching the landfills, which is regarded as a last resort in waste management.

The concept of waste recovery is usually applied for the generation of electricity or power from waste. It can be regarded as one of the sustainable forms in reusing the materials productively and thereby increasing the life span. The high calorific value of waste leads to high efficiency in waste to energy technologies. From collection to segregation and disposal done incorrectly leads to failure of such available technologies. Segregation from biodegradable and non-biodegradable is necessary for the effective application of waste recovery methods. From food waste (biodegradable) to plastics (non-biodegradable) have different methods for waste recovery. *For e.g.* Coffee husks are pyrolyzed and the heat generated is used to dry coffee berries and the bio-product produced which is regarded as biochar is used as a fertilizer in soil (UNFCCC). Before we further move into the depth of waste recovery; it is important to understand the positive correlation of waste and income levels of different countries. High, Middle-and Low-income countries vary too much in terms of generation to disposal. Alone High-income countries generate 34% of the total waste *i.e.,* 683 million tonnes Daily per capita increase of waste for low- and middle-income countries will be 40 % or more for which high income countries is predicted is 19% (world bank report "What a Waste 2.0"). It is due to the fact that the incremental change in the incomes for low- and middle- income countries is faster than the high-income countries. The information is relevant so as to draw a parallel line between the quantity, composition, segregation and disposal options adapted by different countries. As already mentioned, waste recovery method can create miracle only if there is coherent understanding of waste characterization (which will be studied deeply under section 1.3) of these countries. Table 1 mentions about the total waste in percentage from collection to disposal preferred by different income level countries.

Table 1: Data of waste characterization (in percentage) of different income level countries Waste Characterization

Countries	High income (%)	Upper Middle income (%)	Low Middle income (%)	Low income (%)
Generation	34	32	29	5
Waste Composition				
1. Food & Green	32	54	53	56
2. Metal	6	2	2	2
3. Paper & Cardboard	25	12	12.5	7
4. Rubber & Leather	4	1	<1	-
5. Glass	5	4	3	1
6. Plastic	13	11	11	6.4
7. Wood	4	1	1	<1
8. Others	11	15	17	27
Waste collection				
1.Urban	100	85	71	48
2. Rural	90	45	33	26
Waste Disposal				
1.Open dump	2	30	66	93
2. Composting	6	2	10	0.3
3. Recycling	29	4	6	3.7
4. Landfill	39	54	18	3
5. Incineration	22	10	-	-

Source: What a waste 2.0, (Kaza *et al*, 2018, World Bank Document)

After knowing the type and quantity of waste generated by different countries; it becomes easy to understand the best suitable waste recovery method for maximum resource production. The Table 2 compares between the common methods employed for waste recovery for any waste.

Table 2: Methods and example of techniques in waste recovery

Properties	Thermal Conversion	Bio-chemical Conversion	Thermochemical Conversion	Electrochemical Conversion
Definition	Degradation of waste at high temperatures.	Decomposition of waste mainly organic by microbial action to produce useful end products.	Degradation of waste at high temperature with additional application of chemical reactions.	Decomposition of waste in presence of microbes which are used as a charge conducting medium.
Operating Conditions	Aerobic condition with complete oxidation.	Anaerobic, high moisture content of waste and biodegradable.	High heat conditions, less oxygen, low moisture content of waste and non-biodegradable.	Oxidation- Reduction (redox) reactions with immobilised microbial cells.
End Products	Steam, purified flue gas (oxides of nitrogen, sulphur and carbon) and ash.	Methane gas and alcohol.	Heat, Gas, oil and char.	Biohydrogen.
Technologies	Incineration	Bio methanation	Pyrolysis, Gasification, Combustion	Still at nascent level.

(Source: Matsakas *et al*. 2017; Singh *et al*. 2017 and Puna *et.al,* 2010)

3.3 Monitoring, Characterization and Analysis of Waste

Management of waste is a huge issue. Improvement in performance in waste management is a dispensable part so as to understand the nature and amount of waste collected on-site. Monitoring, characterisation and analysis of waste are the backbones of waste management. Characterisation of waste is defined as the scientific approach with proper statistical portrait displaying amount, nature, composition and disposal methods for a waste (Bharti *et al.*, 2019; Kumar *et al.*, 2009; Wainaina *et al.*, 2020)which are facing a serious pollution problem due to the generation of huge quantities of solid waste. This paper presents an assessment of the existing situation of municipal solid waste management (MSWM. It aims at accessing and evaluating the performance of the waste management, provide credible and reliable information about the nature, composition, collection, segregation, generation and disposal (BOMA, Quebec task force committee report- Guide to efficient waste characterization). It provides alternative environment-friendly solutions rather than adopting incineration and landfill. Characterization even helps an individual to optimise the operating costs, diversion rates (*i.e.,* Mass of waste excluding contaminants divided by the total waste) and decrease in contamination. Now the question arises; how can we carry out characterization for any waste? Waste characterization needs a valid and dependable methodology. Fig 2. Summarises the steps to be followed for characterization of any waste.

Define Waste

Segregation of waste on the basis of its composition, nature, etc. For e.g. consumer products (paper, packaging materials), Hazardous waste (Paints, gas cylinders) and construction and demolition waste.

Sample Period and Duration

It is the first important step further going in depth about waste characterization. Sample period and duration can be selected as per the need for a waste characterization. This further adds to the credibility of the data and results obtained statistically in characterization.

Identification of sample waste

One of the most important tasks in waste characterization. The identification gets easy when done on-site. Detailed information of collection site, date, day and time of collection, differentiation into different categories. This even helps in tracking the amount of waste generated, its quantity and composition.

Waste segregation is done according to its nature, composition. It must comprise of the checklist like weight, date, day and time of collection and disposal of the same waste. Checklist must even contain the disposal options different waste undergoes to

Statistical Analysis

Separate and detailed information on waste generated, collected, disposed, composition, etc.

Fig. 2: Stepwise methodology for characterization of any waste
(*Source*: Methodology for the analysis of Solid waste, EU, 2004)

Taking example explaining the waste characterization statistically. A company conducts two working days and one weekend for collection. The waste collected comprises of Organic, Recyclable (non-hazardous) and hazardous waste. Table 3. Showcase the total amount of waste collected each five day with an individual weight of different non-hazardous and hazardous materials.

Table 3. Illustration of characterization of waste by example (Source: Municipal Solid Waste, Ministry of Urban Development)

Sampling Period

	Monday			Thursday			Sunday			Total (Kg)			Sum (Kg)
Types of waste bins Composition of waste	O/B	R	H	O/B	R	H	O/B	R	H	O/B	R	H	
Paper	200	350	0	210	330	0	150	280	0	560	960	0	1520
Corrugated cardboard	0	520	30	0	220	10	0	180	5	0	920	45	965
Food waste	900	0	0	860	0	0	1020	0	0	2780	0	0	2780
Aluminium cans	0	12	3	0	6	1	0	33	10	0	51	14	65
Tins	0	5	1	0	8	4	0	7	3	0	20	8	28
Batteries, cells and	0	0	20	0	0	14	0	0	25	0	0	59	59
harmful chemicals													
Glass	0	2	0	0	5	0	0	5	0	0	12	0	12
Plastics	0	40	10	0	32	3	0	50	6	0	122	19	141
Total (Kg)	1100	929	64	1070	601	32	1170	555	49	3340	2085	145	5570

O/B: Organic/Biodegradable, R: Recyclable, H: Hazardous

From the above table, inferences regarding the type and composition of the waste collected during the sampling period is mentioned in ***Fig. 3*** and ***Fig. 4***.

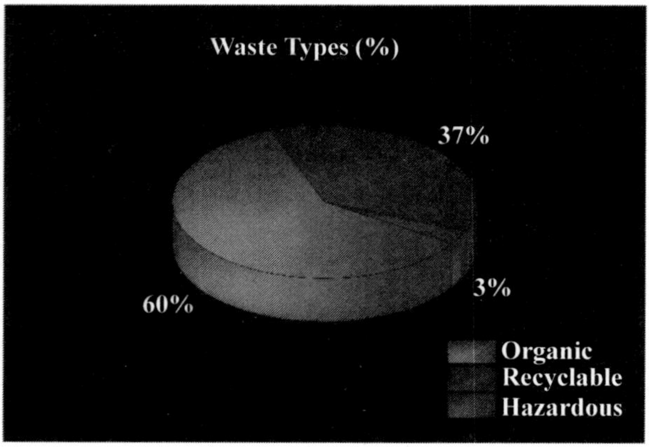

Fig. 3: Type of waste collected (%)

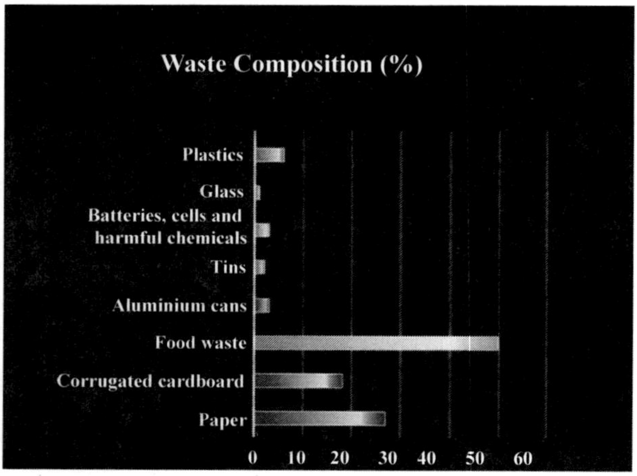

Fig. 4: Composition of waste (%)

3.3.1 Sampling

It is one of the basic procedures in the waste management process. Sampling is a physical collection that represents the bulk waste. This adds to the systematic collection and handling of the whole waste so as to retains its indigenous form and composition. There is a stepwise procedure to carry out an effective sampling so as it does not hinder the physical and chemical properties of waste. Fig. 5. showcases the effective stepwise methodology adapted for the sampling of any waste.

1. Waste Segregation

i. Heterogenous composition of waste: Dry and Wet waste

ii. Various sources of waste: MSW, HW, C&D, etc

iii. Different physical and chemical form

iv. Recycle, Reuse, WTE and landfill options

II. Waste Strategy

i. Objective: Quantitative such as number of samples to be analysed on the basis of the aim of the sampling

ii. Qualitative such as target compounds to be examined

III. Sampling Methodology On basis of Collection

i. Grab Sampling: Sampling at a single time and place with the composition of the source is known.

ii The Variability is obtained for a particular sample but it is a costly method, ii. Composite Sampling: Small fraction of samples form a large one. Hence, reducing the sample number. More precision and accuracy were obtained. The average concentration value of a sample is obtained.

Types of Sampling

i. Random

- Simple Random: Random selection of the samples takes place. Mainly used in homogenous sampling. The sample size and data analysis are easy to perform.
- Stratified Random: Sampling for the segregated or non-uniform area by dividing into non¬overlapping subgroups. Spatial and temporal sampling can take place.
- Systematic/ Grid Random: Sampling was done at a specific or fixed space and time intervals.

ii. Authorative

- Samples with known concentration, physical and chemical properties and composition are sampled. No option for random sampling takes place.

IV. Sampling Equipment: It is selected on die basis of type of waste

i. Solid waste: Grain sampler. Sampling tier. Trowel and scoop. Soil auger, waste pile sampler and

ii. Veihmeyer sampler

iii. Liquid waste: Coliwasa, plastics, glass, pond and weighed bottle sampler iiL Gas waste: Head space container with PTFE -lined septa

Fig. 5: Stepwise illustration in sampling of any waste (Source: Methodology for the analysis of Solid waste, EU, 2004)

Waste Characterisation: The waste characterisation is divided in the following parts (Recycling and Disposal of Municipal Solid Waste in Low and Middle-Income Countries Perspectives for municipal managers and environment agencies 2012):

Quantity and composition of materials in waste (as mentioned above)

1. Physical, Chemical and Biological waste

We will further discuss the 2nd part *i.e.,* Physical, Chemical and Biological methods.

a) Physical Methods: The need for physical recycling lies in the selection and operation of equipment and facilities, in assessing feasibility, resources and energy recovery, and in the analysis and design of disposal facilities. They include parameters such as the following:

I. Density: the total mass and volume of water content in the waste. It helps in the selection of collection equipment. Often, no distinction has been made between uncompacted or compacted densities. 20-25% the density is increased in the transportation step. It is calculated as

$$\text{Density} = \frac{\text{Mass of A} + \text{Mass of B}}{\text{Density of A} + \text{Density of B}}$$; A and B are masses of two different wastes

II. Particle Size and Distribution: The materials are recovered, especially by mechanical means such as trommel screens and magnetic separators. When the waste heterogeneity is difficult it is helpful in those cases. It improves the rate of reaction. The size of a waste component may be defined by one or more of the following measures:

$$S_c = 1$$

$$S_c = \left(\frac{l + w}{2}\right)$$

$$S_c = \left(\frac{l + w + h}{3}\right)$$

$$S_c = (l \times w \times h)\frac{1}{3}$$

Where S_c = size of the component, in (mm)

L = length, in (mm)

W = width, in (mm)

H= height, in (mm)

$$S_c = (l \times w)\frac{1}{2}$$

III. Moisture content: the weight of moisture per unit weight of wet or dry material. It is expressed in %age. The soil is dried at a constant temperature of 100-105°C. The moisture content influences the heat of combustion and the organic matter. Field capacity can be retained in a waste sample subject to a downward pull of gravity. The field capacity of the waste material is of critical importance in determining the leachate formation in a landfill. The water in excess of the field capacity will be released as a leachate. The field capacity varies according to the degree of pressure applied and the state of waste decomposition.

$$\text{Moisture content } (\%) = \left(\frac{a-b}{a}\right)100$$

where

a = initial weight of sample as delivered

b = weight of sample after drying

Field Capacity (FC) (%) = 0.6 – [0.55 * (W/4500+W)]

Where

W = Overburden weight calculated at mid-height of the waste in lift (kg).

b) Chemical Methods: It is important in evaluating alternative treatment and recovery option such as the incineration process. The waste is a mixture of combustible and non-combustible semi-moist materials. It checks on the leachate and ground water qualities. Information about trace elements are known.

I. Ultimate analysis: It is used in determining the percentage of C (carbon), H (hydrogen), O (oxygen), N (nitrogen), S (sulfur), and ash. The emission of chlorinated compounds during combustion is a concern, hence the determination of halogens is often included in the final analysis. The results of the ultimate analysis are used to characterize the chemical composition of the organic matter in MSW. It is also used to determine the appropriate mix of waste materials to achieve appropriate C / N ratios for biological conversion processes.

II. Proximate analysis: It is more specific than ultimate analysis. Determines the moisture, ash content (non-combustible part), Volatile matter and fixed carbon component (combustible part) in a waste.

Fixed carbon (%) = 100(%) – moisture (%) – ash (%) – volatile matter (%).

III. Heat of combustion: It measures the energy released when burned. It can be calculated by bomb scale calorimeter or through ultimate analysis. A heating value of 11.6 ×10 J/kg is required to sustain combustion. The heat of combustion decreases with increase of moisture and organic matter content.

IV. Heating value of other wastes: When the organic fraction of municipal solid waste is used as a feedstock for the production of biological conversion products such as compost, methane, and ethanol, information on nutrients and essential elements in the waste materials is important in terms of microbial nutrient balance and in assessing the final uses of the residual materials after biological conversion, which is calculated using following equation:

$$\text{Btu / lb} = 145.5C + 620\left(H- \frac{1}{8}O\right) = 41S$$

Where,

C = carbon, percent

H = hydrogen, percent

O = oxygen, percent

S = sulphur, percent

c) Biological Method: A characteristic of the organic part of MSW is that almost all organic components can be biologically converted into relatively inert organic and inorganic gases and solids. Odor production and generation of flies are also related to the rotting nature of organic matter present in MSW (for example, food waste). Volatile Solids (VS) content, which is determined by ignition at 550°C, is often used as a measure of the biodegradability of the organic fraction of MSW. The use of VS in describing the biological decomposition of the organic fraction of municipal solid waste is misleading, as some MSW organic components are highly volatile but low in biodegradability (eg, newsprint and some vegetable trimmings). Alternatively, the lignin content in the waste can be used to estimate the biodegradable fraction, using the following relationship:

BF = 0.83 – 0.028 LC

Where

BF= biodegradable fraction expressed on a volatile solids (VS) basis 0.83 = empirical constant

0.028 = empirical constant

LC = lignin content of VS expressed as a percent of dry weight

3.3.2 Waste Analysis

The ultimate objective of analysis of waste in its management is to involve comprehensive evaluation, assessment of goals and scopes, stringent applicability of proper framework and policy. In addition, the mass balance *i.e.,* the sum of input equals to the sum of outputs and increasing and decreasing stock; keep a check on the credibility of the data in terms of its quality and availability. It helps in decision making process. These methodologies adopted during the analysis helps in practical, cost-effective outlook to future sustainability. First use of methodology was undertaken in economic assessment of waste studies in 1965. The use of the computer based was done in 1980's. The immediate and long-term goal of the use of the methods in assessing the waste is synergistic relationship among the

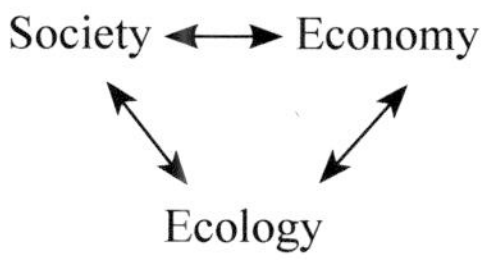

(Allesch and Brunner 2014).

Life cycle assessment (LCA), Cost benefit analysis (CBA) and Multi-criteria decision making (MCDM) are the widely used methods for analysing waste. There are two kinds of methods; one is procedural and other is analytical. Procedural is based on a procedure and deals with society and decision making. While analytical includes the technical aspect. Procedural being greater than analytical hence the latter comes under procedural method. All the methods such as LCA, CBA, MCDM, RA *etc* are examples of analytical while Environment impact assessment (EIA) and Strategic environmental assessment (SEA) come under procedural **(Finnveden** ***et al.*** **2007)**. Now we will further discuss about different methodologies in detail.

1. Life Cycle Assessment (LCA): The name suggests the approach of "Cradle to Grave" that is the assessment of the product from being a raw material up to its end of life that is the disposal and analysing the potential impacts on the environment because of them (Allesch and Brunner 2014). It identify and quantify the energy, resources used and the waste produced as the result and thereby the evaluation of the product or process with potential effect on the environment (Borodin *et al.* 2015). Life cycle costing (LCC) deals with the costing of the product whole of its life time. According to ISO14040 (2006) LCA have four phases:

1. Inventory analysis
2. Inputs and outputs compilation
3. Life cycle impact assessment (LCIA) and
4. Interpretation. LCIA deals with the components such as resource depletion, threat to human health, ecology and analysing the valuation part of it. ISO standards for LCA have explained

LCA taking the following steps (Kiker *et al.* 2005)

I. Goal and scope: Potential impacts lead to the identification of the possible different alternatives to combat the impacts giving resourceful results. In case of SWM; waste being heterogenous in nature leads to the selection of various alternatives for its disposal like composting, recycling, landfill and incineration.

II. Life cycle inventory: Evaluation of the best fit alternatives on the basis of its impact on the environment, social and economy. In case of landfill the possible factors are considered before the approval that is gas generation, leachate generation with and without treatment, presence and storage of toxic substance, construction and operation and carbon sequestration. The primary (face to face interview, site location, *etc.*) and secondary data (published papers and documents and reports) is collected for the inventory.

III. Impact assessment: Potential and possible impacts are assessed. Climate change, ozone depletion, eutrophication, *etc.*

IV. Interpretation: Economic analysis that provides the eco-friendly and cost-effective approach whereas sensitivity analysis assesses one parameter at a time and thereby checking all the plausible impacts. This one at a time assessment is termed as a One factor at a time (OAT).

The objective of LCA is to substantially reduce the environmental impacts. Assess the flow of energy, quantifies the materials used and emissions produced. Practical ideas for the improvement in any step-in process of waste management from collection to disposal. The only drawback using LCA methodology in waste analysis is that a greater number of assumptions lead to more divergence and distraction from the actual results.

2. Cost Benefit Analysis (CBA): The methodology with implications on the socio-economic aspect. It is one of the types of analytical method which evaluates the benefits and costs due to a project. Being a material cycle society, it is important to analyse the waste on all the fronts covering the socio-economic benefits and losses. Just not the socio-economic but spatial, temporal, environmental fronts all are covered. CBA has been integrated with EIA (procedural) that calculates the cash flow *i.e.*, in form of Net Present Value (NPV). CBA helps to support the decision-making process. It is calculated in form of cost/ benefit ratio. If C/B >1; then the decision is favourable while if the *vice-versa* occurs hence it is unfavourable and the project will be rejected. In SWM; the CBA was analysed using a conventional approach that is called as the bottom-up. In this the items are arranged in a hierarchy and the cost estimation occurs at every stage. Its limitation is that we are not able to possibly calculate the cash flow of every item in a category within the credible boundary. The strength of the method is that if the data is unavailable the analysis can be done by estimating the cost. The latest and most accepted method is the econometric method. This method is based on the multivariate regression analysis. Its main aim is to establish a rational relationship and evaluating the damage, financial cost. Usually for better results and macro aspects both bottom-up and econometric methods are used (Weng and Fujiwara 2011). Cost effectiveness analysis (CEA) an integral part of CBA. Evaluation of the cost so as to achieve the goal. CEA has more weightage than CBA as we need no quantification and monetization. The limitation of CBA method is that the intangible is not accounted and nearly impossible to analyse (Allesch and Brunner 2014).

3. Multi-Criteria Decision Making (MCDM): It is multi-disciplinary in nature. It covers all the multi-dimensional conflicting criteria with large number of variables with information available and aim to study. The three pillars of MCDM is economic, social and environmental. It plays an important role in SWM in making decision. The aggregated and incomparable values are easily resolved by this method. The incommensurable variables with several possible alternatives (Coelho *et al.* 2016). The only limitation is that MCDM methodology being highly theoretical and subjective it is hard to infer and interpretation is tough (Allesch and Brunner 2014). There are two main classifications of MCDM which is Multi-attribute decision making (MADM) and Multi-objective decision making (MODM).

`Multi-attribute decision making (MADM)	Multi-objective decision making (MODM)
Selection of ranks or problems	Optimises the ranks or problems
The opted criteria prioritise the rank or problem.	The criteria selected in MADM method only that rank is optimised further.

Discrete decisions with limited and pre-determined alternatives.	Continuous flow with unlimited and no pre-determined alternatives.
Mathematical operation like use of matrices for the calculation of MADM is used considering the evaluation of criteria for the alternative.	Mathematical operation like use of vector method is considered for the evaluation.
MADM application in SWM is in areas like facility location, selection of treatment technologies.	MODM application in SWM in areas like collection, routing, planning and allocation of resources.

MCDM method framework is divided in systematic phases for its convenience to understand and interpret easily. The phases are as follows (Coelho *et al.* 2016)

a) Goal and scope: The goal concludes the selection of framework and scope defines structure of proposed assessment. Focuses on problem selection and its steps to rectify. Status, performance of criteria both qualitative and quantitative assessment is conducted. It measures directly the performance of the project/process. MCDM is directly proportional to the framework adoption *i.e.,* the any inferences concluded by the MCDM method it will reflect in the adoption or rejection of the framework, policy *etc.*

b) Criteria weights: It depends on the two-minute methods which are critical factors in MCDM methodology. First one is Opinion based approach that deals with the subjective handling and opinions of various stakeholders and experts emphasise the decision making. The second one is the data based that deals with the mathematical model.

c) Data normalisation: It is the normalisation by comparing the different values/ scales/ units for different criterions. It helps in the linear transformation of raw data by translation and expansion. Not all the criteria require normalisation. Indicators with same units/scales does not require normalisation and the requires normalisation in *vice-versa* case.

d) Sensitivity analysis: The characteristics which is insensitive to small changes in the input parameter. Even the variations in the output parameters are evaluated which is impacted by the input. The consistency of final decision and priority areas are ensured. It is of two types: One factor at a time (OAT) and Variance based method (VBM). OAT deals with the variation in the output and VBM deals with the total output variance in accordance to the input.

In SWM the MCDM method is used as follows:

Method	MCDM
Goal and scope	Waste management, strategy, facility location, transportation, collection and routing.
Criteria	Environmental, Social and Economic
Weighting method	Opinion and data based
Normalisation	Linear
Sensitivity analysis	VBM

Following are the summarized other methodologies used in SWM for the waste analysis3

Methodology	Description
Risk Assessment	Assess the organisation performance, process and individuals. It is accountable for the comprehensive or fully defined risks pertained.
Bench marking	It adds to the continuous evaluation and comparing simultaneously different products, processes, *etc.* Help identify the loopholes with effective solutions.
Cost effectiveness analysis	Intangible effects dealt. For a process full evaluation on basis of its advantages and disadvantages.
Eco-efficiency analysis	Environmental and economic aspect go hand in hand. Cost benefit of product is analysed on impacts, effects basis.
Environmental impact assessment	Baseline study which evaluates the economic, social and environmental aspects.

As discussed above the strengths of the methodology in analysis of waste. There are few limitations that we have come across. Methodology leads to the uncertain conclusion and wrong method selection. Cost determination and its effectiveness is tough which requires lots of improvisation for the long-term perspective. It is imperative to understand and have full knowledge about the product/ material as a waste. It helps in evaluating the possible potential emissions released by the product. It is essential to set a time frame which will help in evaluating and comparing the future emissions which are more or less dangerous in future than the present. The best solution available is the integrated approach. Combing two or more methods like CBA with EIA or LCA with GIS *etc*. They are beneficial in management and planning, policy making, increase the knowledge and provide common grounds for organisers and groups for discussion.

3.4. Recyclable Materials

1.3 billion tonnes of waste (2011) is increasing day by day and estimated to reach up to 2.2 billion tonnes by 2025 according to the reports by World Bank (Coelho *et al*. 2016). According to USEPA data; 262.4 million tons of MSW is produced in year 2015 and out of this highest percentage of materials in MSW is paper and cardboards with 25.9 % but the figure was reduced from 84.8million tons (2005) to 68.1 million tons (2015) due to digitization. 49.7% of the paper and majorly corrugated material (28.9 million tons) were recycled in 2015. In case of glass 3.3 % (3 million tons), plastics 3.4 % and metals 9% (USEPA Facts and figures material waste and recycling). Recycling being one of the convenient and eco-friendly methods of all approached for the waste management. In simple terms it is defined as the process of collecting, segregating and reprocess these waste materials to form brand new materials (Bom *et al.* 2017). Recycling one of the process that helps in the production of the environmentally sound and sustainable products. The replacement of materials by the recycling materials; cut down Greenhouse Gas (GHG's) emissions by 20%. The study in Hongkong; the replacement of recyclable constructions such as bricks, plastics, asphalts by concrete building materials GHG's emissions reduced up to 6 to 17%. Same study in Taiwan on replacement of construction waste by recyclable construction waste by recyclable construction waste reduction of GHG's observed were 16-23 % observed (Jimenez *et al.* 2018). The main goal of recycling is to reduce the burden on the environment in terms of resources, energy and waste production. It minimises pollution and ecological degradation (Waste section 6(1); National Waste Management Strategy). In this section we will learn step-wise recycling process for various recyclable materials. Recycling has two stages: Material separation and material conversion. In separation the mixed waste is separated on basis of its physical and chemical properties. In conversion; the materials obtained after the separation process is used as secondary raw material (Muttamara *et al.* 1993).

3.4.1. Recycling of Paper and Carboard

Recycling of paper and cardboard waste are easy process. The steps in both the recycling steps are similar. The process involves the collection the materials are collected by the door to doorman and further it is sold to the middle man to small business-man to small-man business which first sort and pre-treat and then sells to the recycler. The question of proper material recycling is addressed by making the right choice in printing substrates (high raw material content, chlorine free processes), biodegradable inks, printing press with cleaner and energy efficient and cost- effective technology. Post processing involves recycled paper without glue contamination. Technology such as the digital electrophotography indigo press. It releases 5-20% of waste paper and 0.2-1.1 g per impression as chemical waste. The other one being Sheet fed-offset press produces 35-75% of paper waste and 3.2-7.3g per impression of chemical waste. Even emission of VOC of 74mg/cubic meter (in breathing area of press) and 484 mg/cubic meter (in conventional press) with n-hexane content. If the volume of the digitally product increases; the paper waste recyclability becomes very important. As materials can be reused from cradle to cradle (*i.e.* back to the original form) (Mirkovic *et al.* 2015). Materials like newspaper, magazines, cardboard and bottles are separated at source. The larger pieces are removed and further sent for further processing. Following steps are mentioned required for the recycling process for both paper and cardboard (Haupt *et al.* 2018).

Recycling Steps	Paper	Cardboard
Collection	Waste collected from collection point i.e., trash bins, paper store or scrapyards etc. The waste is measured, graded for quality and auled to the recycling mills	
Sorting	Separation of the paper and cardboard on the basis of materials to be recovered and the end product formation	
Shredding and Pulping	The waste is cut into smaller pieces by passing through shredder. The waste is then placed on the transporter belt and passed through pulper. In pulper the addition of water and chemicals breakdown the waste paper to the pulp. The non-fibrous part is discarded likewise staples, paper, etc. and the fibrous part it goes in for deinking process and is filtered and screened again to remove the contaminants such as glue, inks, staples etc. The pulp is mixed with the new pulp and form a firm good grade paper. The cleaner pulp is then placed in machine to centrifuge and remove more debris from paper pulp	In case of cardboard; the above process is same and the non-fibrous discarded part is fed for the thermal utilisation. In test liner (two piles of paper); composition difference in each layer. To increase the strength and stiffness starch treatment and surface treatment (by size press) is conducted. The treatment helps to receive high quality corrugated fibre base. Size press comprising of two revolving rubber rolls pressed together. The starch solution is present at the end of the roll and hence the paper absorbs and passed through dryers to remove excess of water
Filtering, deinking and decontamination	The slurry using centrifuge is then is helpful in removing further contaminants such as plastics, metal scraps, etc. The light material float above and heavy material present downward is removed. For further removing the dye and ink the floatation device made up of chemicals and air bubbles are employed. Hydrogen peroxide is further applied to bleach the pulp for cleaning and ensure its originality in case of paper	
Application	The pulp is passed through hot and dry conveyor belts of cylindrical form. The pressing removes excess of water and the heated cylinder rolls form flattened sheets. The recycled paper is rolled and sent to newspaper printing, wrapping papers, printing papers, etc.	The cardboard is passed through the slitter winder roll. The big sheets of cardboards are sent to the corrugated industry. The secondary fibres (derived from the primary cellulose) are used as an insulation material

3.4.2. Recycling of Plastic waste, tires and electronic waste

According to USEPA 2015; 3.4% of plastic is recycled. Recycling of plastic involves different approaches for recycling. The post-industrial has no organic residue and the composition is the same. The mixed plastics are recycled differently in comparison to the other plastics. High density polyethylene (HDPE), low density polyethylene (LDPE), polypropylene (PP) and polyethylene terephthalate (PET) are recycled differently (Drzyzga and Prieto 2018).

In plastic recycling the waste is first sorted; the plastic bottle waste majorly PET bottles are recycled. The wire is removed around the PET bottle. The bottles are then passed through a metal separator that separates the aluminium-containing labels. The 12mm size shreds the bottle into pieces which is referred as PET flakes. The label residues are removed from flakes by the air separator. The residue is fed to the municipal incinerator. Bottle caps are removed with the help of a process called the sink-float. Caps float on the surface of the water because it is light and heavy flakes sink in the water. Then the caps are separated and sent for recycling. The polyethylene terephthalate (PET) flakes are treated with 50% sodium hydroxide (NaOH), and the flake liquor is roughly heated at a temperature of 200°C. The solution dissolves the PET surface and removes SiOx from the bottle cap. The purified polyethylene terephthalate flakes are washed with water and dried. Hence two types of PET grades are produced *i.e.,* amorphous and bottle grade. Amorphous PET produces trays and other packaging materials and PET fibers. While bottle-grade is used as a substitution to the other bottles produced. Mixed plastics such as High-density polyethylene (HDPE), low density polyethylene (LDPE), polypropylene (PP) and polystyrene (PS) *etc* are present. HDPE have low density than PET and can be separated by the sink float method. HDPE and others are even separated by the near infrared radiation technique (NIR). Plastic must not be dark so as to absorb the radiation. Sorted plastics is shredded, melted to form refine polymer. The polymer is cooled and form pellets and used in manufacturing. The high-quality grade plastics substitutes HDPE, LDPE, PS *etc* use by 90-70%. The low-grade quality substituted by wood and concrete by 50% and pallets, park bench *etc* by 7-14%. Beverage cartons is also one of the mixed plastic waste examples. It comprises of 75% paper, 21% plastics and 4% aluminium. The dissolution process takes place. In this process the sterilization of the carton takes place with steam. 97% of the carton is recycled by cardboard recycling and rest of it are collected and dried and fed to thermal utilisation process. The carton fed through the cardboard recycling is ended up using as corrugated board production mainly the liner board is produced (Haupt *et al.* 2018).

The tire recycling is that it cuts the tire in to smaller pieces and further process it for additional use. Reduction in the volume and make the storage easier. It is even used as the tire derived fuel (TDF). Further more valuable product like crumb rubber, steel and fibre, *etc* are also derived. The following steps are involved in the recycling of tire waste. 1. Shredding of tires: speed rotatory shears or hammermills shred the tires in small parts. It helps to reduce the volume. These are called as tire chips. The chips are further processed to increase the quality. Mechanical grinding method is employed to remove steel and fibre. Use of serrated grinders produce the crumb rubber. 2. Cryogenic reduction:

Freezing of the material at -325 Fahrenheit under liquid nitrogen. Being brittle in nature and hence fed to the cooled closed loop hammermill. Under this process the crumb rubber formed won't be altered and crumb rubber formed is of appropriate size and form. Majorly tire waste is sent to the landfill and is widely used as the one of the recycling processes. But due to environmental constraints like damage to the clay, leaching *etc* this practice is no longer practiced. 3. Tire derived fuel: The scrap tires are burned (> 3000° Fahrenheit) wholly or fed to TDF. It can be used as a fuel due to its high energy than coal (13200 btu/lb) and less sulphur emission with air emissions controllable with the standard norms. TDF is used in cement kilns, other facilities and paper mills. Cement kilns: High temperature requirement and burning of exhaust gases for long time. When the steel heats at high temperatures in tire shreds the steel produces iron as a supplement with other materials. More energy than coal and btu value higher; with nitrogen, sulphur and ash contents being lower. It is preference in cement kilns is marginal due to difficulty in the air permission and its compliance with the standard air quality norms. It is time consuming and a costly affair. Paper mills: Large amount of wood scarps are produced. The hog boilers work with coal and produces steam. To maintain the working of the boiler good quality fuel is required and hence TDF acts as a competitive option. More btu and low moisture content acts as an advantage but again the permit is an issue same as the cement kiln (Owen 1998). Waste tyre rubber is also used as an oil absorbent. The graft copolymerization method with 4-tetrabutylstyrene (monomer) with waste tyre is one such method for recycling of tire waste. The Divinylbenzene and benzoyl peroxide is used as a crosslinker and an initiator respectively in the process. The increase of the ratio among waste tyre and butyl styrene and Divinyl benzene lead to increase in the oil absorption (Wu and Zhou 2009). Usually, the waste of tyre is pyrolyzed and the pyrolyzed residue has further no use except to dump. But in recent development the Pyrolyzed residue is treated with strong acids and a diacid combination of HCl and HF under ultrasonic waves. The carbon black called as pyrolytic carbon black (PCB) is formed. The properties of PCB are equivalent to the commercial carbon black (Zhang *et al.* 2018).

E-waste one of the growing wastes in terms of volume and impacting the environment. E-waste contain up to 60 metals and precious metals like Cu, Au, Ag Pa and Pt. The recovery helps to reduce the pressure on extracting more of these precious metals, less burden on landfill and environment. So far only 15 % e-waste is recycled. E-waste is defined as the waste discarded with no intension to reuse by the consumers; including all the electrical items. Two ways of recycling e-waste are through formal and informal sector. Formal sector around 60-70% of e-waste is sent to landfill (Mandpe *et al.*, 2019a). In developed countries the e-waste is collected and sent to the different e-waste processing units, and somewhere it gets mixed with the municipal solid waste and sent without sorting to landfills. In developed countries informal recycling of e-waste is widely practiced and then sent to landfill or incinerators. The recycling of e-waste accounts for the following (Kumar *et al.* 2017):

1. Pre-processing- Dismantling electronic devices and separating metals, plastics, *etc.* Few materials cannot be separated by hand and thus sent for shredding. Certain metals and plastics are removed using magnetic separators and gravity separation.

2. Dismantling- Manually separating and the materials non-separable are sent to further processing. Its use is removal of hazardous waste and no dust with only high-grade materials sent for further processing. Disadvantage is high capital cost for labours, public health issues, time-consuming, *etc.*

3. Shredding- Metal shredders, hammer mills and knife mills help reduce particle size by crushing and grinding electronic waste. Faster automated system with reduced public health burden, lower volume, and thus easier portability. Higher capital investment, high dust, material loss up to 40%, lower grade quality is obtained.

4. Mechanical separation- Dry and wet e-waste is separated. The wet waste is separated with floatation and gravity separators. Fe, Steel and other ferromagnetic and rare materials are separated. Precious metals like Cu, Ag and Au are separated using density separator technology such as air tables, air cyclones and centrifugal separators. Eddy current separators employed for Al recovery. Nowadays various sensors are employed for the separation such as optical and infrared sensors for glass and plastic removal. The negative part is that the huge capital investment and its

applicability only in huge recycling units and not suitable for small. Huge dust in dry systems and moisture control for the wet system.

5. End-processing- Recovery and purification of metals such as Ag, Cu, Au, Pa and Pt, *etc.* Process involves Pyro-metallurgy, Hydro-metallurgy and Bio-metallurgy.
6. Pyro-metallurgy- Melting of materials at high temperature to derive mixture of purified metals called as electrorefining. The recovered metals like Al, Cu, Au, Ag and Pa are usually oxidised. Faster reaction rate at high temperature helps to separate valuables and waste. High energy requirements, with the generation of dioxins, furans and volatile minerals, and risks to the environment and public health. Partial purity of the precious metals and plastic recovery is not possible.
7. Hydrometallurgy- The leaching of the concentrates during the pre-treatment and recovery of materials with help of chemicals during the process is called as hydrometallurgy. Electro-winning is the device which helps in hydro-metallurgy process long with some chemicals which enhance the process. Less energy intensive and easily accessible with more accuracy. High toxicity with more no of chemicals involved, slow and time-consuming, high cost and requirement of fine grinding particles.
8. Bio-metallurgy- The use of microbes which help in leaching of the precious metals. It is an eco-friendly approach. Usually, Ferroxidase which is an acidophilic thermophilic bacterium widely used to leach Cu and Au. Purification by electro-winning process. Eco-friendly with low cost and chemical usage. Higher complexity.

The route of transfer of e-waste is through air, food, water and dust. The increased levels of heavy metals and Brominated flame retardants (BFR) are found in humans (mainly the e-waste workers) leading example of occupational health disease. Polychlorinated dibenzo-dioxins and furans found in the high concentration in serum of the non- exposed workers. Even the breast milk samples of mothers were detected with high concentration of hexabromocyclododecanes (HBCD's). Neurodevelopment disorders and foetal perturbations occurrence observed. PAH lead to the decreased cognitive developmental effects. Air samples contain high levels of heavy metals and congers of polychlorinated naphthalene (PCN) and high level of Al, Cu and Fe were found above the permissible limit as per USEPA ambient air quality standards of air. The main reason is burning of e-waste. Ash samples contain Cu, Zn, Pb, Sn as much as 160,00 mg/kg of ash and Br, As and Hg as high as

1500mg/kg in ash samples. Soil contaminated with dioxins and furans, Cu, Pb, Zn and Br due to metal compound burning. Cu, Cd, Pb, Fe, Cr and Ni contamination in marine due to dumping and burning. Food chain leads to the bioaccumulation and bioconcentration among different trophic levels such as from fishes to humans due to intake of the marine species. Dangerous to aquatic life by effecting the normal homeostasis (Daum *et al.* 2006). Hence it is imperative to recycle the e-waste. Few papers have suggested that the involvement to invest finance for the expansion and new provisions of recycling units in the nations. Stringent application of the latest legislations must be followed. Mobile recycling units must be established in a region or town. Small generation of e-waste in an area can unite together and form a bigger centre/recycling facility (Kumar *et al.* 2017).

3.4.3. Recycling of Glass and Metals

Recycling of glass: The coloured broken glasses are usually not used for producing the new recyclable glass. The glass ends up being a drainage layer and cover material for the landfill. Crushed recycled glass (mixed cullet) is used as a substitute for soil and aggregates and is used in construction materials. The glass cullet if present above the landfill can contaminate the groundwater. The heavy metal detection is also observed due to mixing up of the materials/ chemicals on the glass in to the leachate. There are three different types of recyclable glasses on the basis of geotechnical and chemical properties such as 1. fine recycled glass (<4.75mm) 2. medium recycled glass (<9.5mm) 3. coarse recycled glass (<19mm). Crushed recycled glass can be used as a filler for manufacturing cement, concrete or aggregates in road construction. It helps in production of newly formed glass product. If not washed properly these glass materials have biological and chemical contaminants which cause potential risk to the workers involved in the construction and manufacturing. The leachable chemicals mainly are the ones present on the surface of the glass materials. Usually, the heavy metals associated with the glass are bonded firmly they are dissolved when mixed with water and mainly in the acidic solution and hence can leach out. The testing of the recycled glass is required. Soft drinks, liquor bottles, food *etc*; the mixed colour glass fragments cannot be recycled (Imteaz, Ali, and Arulrajah 2015). Another form of mixed glass waste is Cathode Ray Tube (CRT). CRT is a mixture of waste containing ferrous and non-ferrous metals, glass and plastic. The CRT is present in televisions, laptops, personal computers, etc. The CRT glass contains metal oxides such as Lead oxide (PbO), Barium oxide (BaO) and Strontium oxide (SrO) which cannot be used as recyclable cullet (Maschio *et al.* 2013). The recycling of waste is employed to reuse the glass. The conventional method is to melt the glass so as to reuse or recycle it. The alkali-silica reaction (ASR) is induced with the recycled glass aggregates which has low strength and durability to be used as a concrete. Recent work also shows the replacement of cement and the use of powdered glass (grinding of glass pellets) due to its high content of silica which acts as a pozzolanic material that reacts with portlandite in aqueous cement to form a C-S-H bond that increases the strength and durability of the concrete. Glass is also recycled to produce other valuable materials. Glass once collected is washed, crushed and sieved. The sieving takes places with 10mm and 5mm glass cullets. The 10mm glass cullet produces the replacement of concretes to be used in the pavements, self-compacting concrete and the 5mm glass cullet replacement of sand in the mortar, paving blocks *etc.* Th use of recycled glass in concrete block: The semi-dry mixture is moulded by applying strong compaction. Minimal amount of water is used. The glass contents are different for improving the strength (compressive and flexural). It is reported that the use of 9.4% of glass increases the compressive and flexural strength by 21.1 and 73.3% respectively. <1.18mm of glass was replaced by 20% aggregates improves the compressive strength by 69%, flexural strength by 90%, tensile strength by 47% and resistance to abrasion was improved by 15%. The fitness modulus (FM) plays an important role in deciding the compressive strength. The most suitable for the replacement of the particle sized aggregates by the dry-mixed glass paved blocks are with the FM values ranging from 3.5-4.5. The wet-mixed glass paved can create issues with finding the replacement for aggregates in comparison to the dry-mixed value. If the glass content is less than 20% content it is called as the wet-mixed concrete blocks. This induces no ASR expansion. The pilot scale study identified the correct composition for the use as concretes. Sand, <2.36mm of glass cullet and cement are prepared for applying in the first and 2nd layer and final compaction further the vibrations are applied. Use in self-compact concrete: Glass cullet <5mm, 5-10mm replacing river sand and 5-10mm replacing granite. Increasing the recycled glass cullet content increases slump flow and blocking ratio for concrete mixes. This is due to the water resistance property of the glass cullet and the best use of the LCD slump flow. Use in the architectural mortar: The aesthetics of the glass is retained by the 2.36-5mm and 5-10mm glass cullets for the architectural mortar. White cement, metakaolin, natural sand, recycled glass, water and superplasticizer are used. The ratio of the binder and aggregate is 1:2. The application of the glass in these sectors helps to reduce burden on landfill, reduce the use of the virgin materials for the construction and thereby producing more load on the environment for its management. Increase in strength, durability and resistance against abrasion and acids. Increases the longevity (Ling *et al.* 2013). Production of the green packaging glass is also used. After the screening and crushing process, the glass cullets are melted at 1580°C. Bright molten glass is blown with compressed air to be ready to use glass containers and bottles and bowls are transferred to a cooling furnace. Further the spray is induced to protect from breaking and scratching. This is how a new glass is formed from the recycled glass material (Haupt *et al*. 2018). The recycling of glass is having some disadvantages when it comes to public health. The point of scale recycling is most widely used process and is even encouraged to follow by many countries. It includes the practice of returning the glass bottles to the store from where the glass item is purchased. The study conducted stated that this recyclable practice causes irritation in upper airways and chest symptoms due to fungal exposures (Kennedy *et al*. 2004).

Recycling of Metals: When the metals are produced a lot of scraps and waste chips are formed. The waste and scraps are sent back to the smelters where the recycling and recovery takes place. It is labour intensive, costly and lot of metal is wasted during the process. Two types of materials are recycled one is ferrous and other one is non-ferrous. Mainly Iron and steel who have the magnetic properties are termed as ferrous metals. Metal waste is derived from the vehicles, appliances, machineries *etc*. The scrap waste metal is one of the most widely used raw materials for the steel industry as it reduces the load on extraction of the coal and reduction in the GHG emissions. According to Bureau of International Recycling suggests the following steps involved in recycling of ferrous metals (conventional technique):

1. Sorting: It is the separation of the metal waste with the help of magnetic separators. All the metal and steel are not separated and hence processed further.
2. Shredding: The segregation of the metals from the mixture of metal waste is carried out with the help of rotatory magnetic machinery.
3. Media separation: Electric currents, high pressure air flow and liquid floating system are used.
4. Shearing: Hydraulic machinery exerts enormous pressure and cuts the thick sheet of metal and steel of the railway tracks.
5. Bailing: Compacted into large blocks and transported.

Non-ferrous metals recycling: Metals like Al, Pb, Cu, Zn, Ti, Co, Cr *etc* are regarded as non-ferrous metals. Al is widely recycled of all the others. According to Bureau of International Recycling suggests the following steps involved in recycling of non-ferrous metals (conventional technique):

1. Sorting: Separation of non-ferrous metals from the mixed waste such as paper, plastics, *etc*.

2. Bailing: Compaction of the waste in the large blocks.
3. Shearing: Enormous pressure is applied for cutting the sheets into different sizes.
4. Media separation: Separation of ferrous and non-ferrous wastes occurs using electric currents; high pressure air flow and liquid floating system.
5. Melting: The melted metal is poured in different ingots and allowed to cool. These ingots are further transformed in to sheets or products like tubing.

Metals are categorised in three forms 1. home, revert and run around, 2. prompt industrial,3. postconsumer and obsolete. Home scrap is the one which is generated in the plant itself and only melting is employed. Postconsumer is the one which is derived from various sources such as automobiles, machineries, appliances, old buildings, railway tracks, *etc.* Prompt industrial is the one which is freshly generated and is sold and purchased by the scrap processors or the consumers. The following explains different methods to recycle the metal waste from different metals ferrous and non-ferrous as read above (Kaplan and Ness 1986).

1. Ferrous scrap: Mainly used to produce iron and steel. Oxyacetylene helps in cutting the scraps from the railway, ships and buildings. Shear is the second step which cuts the large pieces of scraps in to smaller form usually 1-1.5m long. Two types of shearing are used, *i.e.,* the alligator like a giant scissors and only the upper part is mobile, and the scissors have two large moving heads that are hydraulically driven and the scrap is cut and protruded behind the anvil. Baled is the step in the process which forms shapes of the metal waste. It consists of hydraulic cylinders which push the scrap waste against the fixed walls. The rotating cylinders have series of cavities which compress the metals in those cavities forming a particular shape. Shredders prepare the light scraps consisting of steel hammers on the rotating drums. The hammers pass over the fixed anvils. Mainly the heavy and large machineries, automobiles are pressed against the hammers and the magnetic separators separate the ferrous and non-metallic materials such as rubbers, plastics, glass and dirt. Further the ferrous metals are sent for detinning process to remove tin. The detinning involves the process of leaching or dissolving the tin in NaOH-NaNO3 solution (detinner). This process removes aluminium also. The tin from the solution is recovered by electrolysis method. The low grade is sent for covering the manholes and reinforcing bars the high grade is sent to make cast of iron and steel.

2. Copper Scrap: Different grades of copper are recycled differently. Grade No1 which is unalloyed copper and is recycled directly by making wire bars. Garde No2 contains copper (Cu) with 94-96% and some impurities. Light refined Cu having 92% is electrolysed, refined grade brass with 61.3% Cu and less than 60% Cu is melted in blast furnace and blown in convertor before electrolytic refining. Conventional and traditional method includes that the shearing of the metal and reduction in size as per requirement, crushing of the thin metal, separation by the magnetic separators, cleaning and greasing. The new method involves the pre-shearing of the large metals. The alligator shear is involved in cutting the large and big metal sheets into small pieces. Three step chopping phases with magnetic separation as the first phase, second includes the sizing and third and final stage of air separation process. The chopping process helps to produce clean metal, removes contamination and insulates the material. The magnetic separators separate and further the media separation occurs. Cryogenic techniques are also used for recycling. It involves the cooling of the scrap at -185 °C it makes the copper malleable and other materials brittle and on crushing the item immediately after cooling we are able to retain Cu by fine sieve. It helps to derive the metals from small motors, transformers, generators, *etc.*

3. Zinc Scrap: Zinc is recycled from drosses, fumes and skimming. Majorly the Zinc (Zn) is derived from the automobiles and appliances. The recovery occurs by mechanical process such as floatation or screening. If the content is too low it is insufficient to recycle and excess of Al and small portion of Zn undergoing skimming method lead in the loss of zinc. Zn recovery decreases with the presence of glass, rubber other trash materials.

4. Magnesium scraps: Magnesium (Mg) is used as a reducing agent for metal production and is the anode for protecting steel in melting furnace. Al and Mg being similar and hence its recycling becomes tough. The reliable method is the use of 5% silver nitrate solution. Applying it to the surface of the metal and the metal containing Mg turns black and the Al remains unaffected.

5. Lead Scraps: Recycling of lead from the batteries: -Removal of the battery from the case and plate along with the posts and connectors. The plates, separators and acid fall on the conveyer belt. The acid is drained into a sump where it is neutralized by ammonia or lime. The air classifier helps to separate the posts and connectors and the battery metal body is broken off. The metal is separated by the help of hydrometallurgy and heavy media techniques. Another method is smelting of the battery after

draining the acid. Lead from the battery 60-65% is recycled while that from solders only 10% is recycled due to its less quantity of lead.

6. Nickel and stainless-steel alloys: Nickel alloys: Ni is employed in gas turbines and in corrosion-oxidation resistance due to its high strength at high temperatures. They are recovered by metal forming and fabrication process. It can be even recovered by electroplating method acting as an anode. This electroplating method is used for recovering pure nickel and only small quantity is retained. Stainless steel: Series from 200, 300 to 400 are found. The 400 series won't be recycled due to 100% chromium in it. In 300 series 18% Cr and 8% Ni other elements like Niobium and molybdenum may be present. They are basically used in aircrafts, kitchen utensils, hospital equipment, *etc.* 200 series have 5-10% manganese and 4% Ni which form alloy named as austenite. Recycling steps include: cleaning or sizing, crushing in ring type or hammer mill then sent for chemical analysis. For superalloys scraps they are degreased with trichloroethylene and screened for fine removal and then further by magnetic separator.

   The superalloy is cut even by plasma torch and cleaned by sand blasting. In stainless steel they are crushed and separated by magnetic separator and further pelletized and sent back to furnace for steel making.

7. Precious metals scraps: They are derived from the plating, dental laboratories, jewellery shops, electrical appliances, *etc*. The recycling of the scraps involves separation of the metals and non-metals as much as possible. He recycling steps involves: shredding of material in uniform size, separation by magnetic separators, screening, air classification and then lastly incinerator to form precious metal concretes.

8. Aluminium Scraps: The conventional process in recycling mainly the aluminium includes melting process as the main step. The conventional method is able to recycle only 55% of the aluminium scrap rest of it is wasted in remelting stage *i.e.,* 20% and post-melting stage losses 25% of the total metal weight. Aluminium and its scrap come under the category of reclaimed raw material and hence when collected they must be sorted on the basis of categories and grades. Before the recycling of the aluminium we must be aware of certain peculiarities such as proper chemical composition and no mechanical impurities. The recycling of the aluminium occurs as follows: 1. segregation of chips, 2. cleaning of the chips, 3. formation of the chips into a final product. The segregation of the chips is performed on the basis of chemical composition and mechanical impurities. Cleaning of the Al chips involves by use of oil

emulsions including coolants and lubricants in machine. The emulsions can be removed by thermal or chemical methods. The thermal methods do not form any carbon deposition on the surface of the chips. Compaction and sintering help in the final product formation. Products formed from the recycled Al chips: 1. Lab results confirm that the AlMg2 and AlCu4 without the use of any hardening substance can form strong bonds due to hot extrusion and press moulding. The chips of 1-5mm diameter and temperature of 500°C is employed which results in the increase of the tensile strength by 105%, relative density by 99% and elongation of the solid material formed is 11%. When tungsten as a hardening substance is used. With addition of 1.5 % of Tungsten slight increase in the hardness observed. Increase of tungsten also increases the yield point and the tensile strength. On addition of graphite (3%); it increases the temperature resistance in such powder-based aluminium (Gronostajski and Matuszak 1999).

3.5. Physical and Chemical Recycling

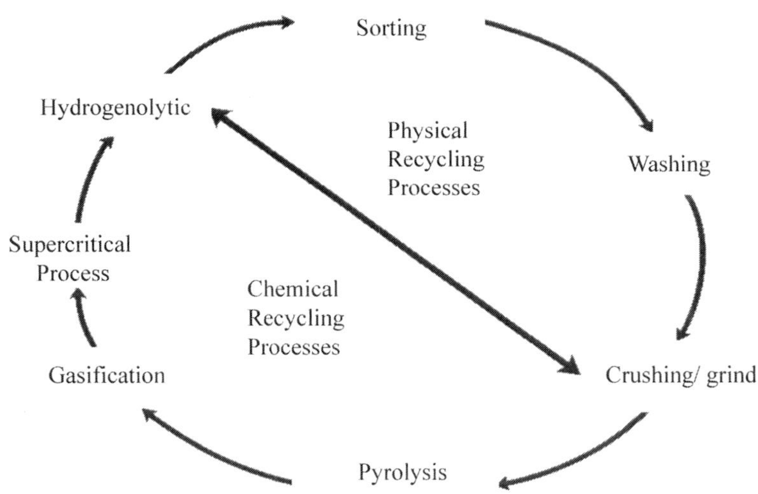

Fig. 6: Indicates the physical and chemical recycling processes (*Source*: Kaya 2016)

Physical recycling is used to upgrade the stage. It includes processes such as crusher, pulveriser, classifier and separator. It evaluates two things: material recovering efficiency and the impact on the environment by the different methods. The advantage of employing physical recycling is the low capital cost but the main disadvantage is that the loss of 10-35% of materials (metals and non-metals) in this process. The physical process includes a crusher, a pulveriser that undergoes electrostatic separation which produces a metallic powder (Zn, Pb, Cu, Al, Au, Ag, *etc.*) and non-metallic. To avoid the dust pollution hurricane, bag filters and air cleaner are used. Chemical recycling is all about transforming the waste (feedstock) into a used product mainly fuels or chemicals (Dhote *et al.*, 2020). The feedstock has sufficient high hydrocarbon content and higher carbon efficiency with no oxygen and less moisture. Pyrolysis, gasification, fluid-catalysed cracking and hydrocracking. Chemical recycling has opened new pathway to reach a sustainable development by means of waste being a precursor adding value to it commercially. The chemically recycled products are more expensive than virgin materials due to the high cost of investment, raw materials, and operation. Now let us now understand the two processes by taking an example of e-waste and PET bottle recycling. Though in 1.4. sections we have come across various combined physical and chemical processes of recycling different recyclable and reusable materials.

E-waste Recycling: Physical recycling- PCBs are the main component in the electric waste. Copper (Cu) is obtained from cable wires with 99% removal efficiency by dry shredding and air classification. The cable wires contain 40% Cu and 60% PVC plastic which are easily separated. Wet gravity separators are used to separate wet PCBs of high density and coarse mixed metals from low density plastics and microplastics. Materials such as glass, ceramics, metals *etc* are removed by mechanical/physical means. Dismantling it systematically removes the components part by part. It is not energy intensive. It is of two types: selective and simultaneously. Selective dismantling is the one with informal separating method. It is usually performed by heating the solder above its melting point (desoldering) using electric blower. Disadvantage is the release of bad smell and black fumes. It is important valuable component as highly valuable components can be extracted. Semi-automatic can be combined with the mechanical recycling. Desoldering process with applying additional forces such as vibration, shearing and impact. Crushing and shredding: This important step helps in the removal of metals from the mixed waste. Though it is causes 40% of metal loss. Dust and harmful gas and dioxins are released during crushing.

For pulverizing ball and disc milling are used to cut the PCBs in small parts. For crushing 10mm torque sheer shredders are used. PCBs with different size particles are separated by methods size, density, magnetic separator, gravity separator, eddy currents and electrostatic. 73.1% organic matter, 4.65% aluminium (Al), 4.55% iron (Fe), 2.67% copper (Cu), 1.06% lead (Pb). Minerals sieved 0.75 μm lower than the particle size will be metal recovery. The shredders reduce dimensions by cutting, tearing, and extruding. Good and reliable quality is obtained with materials of low volume. The double shaft shredder is used for cutting the plastics, waste rubber, wood *etc*. Sieving: For uniform size distribution of the materials sieving is required. Screening and sieving are required due variable shape and size of the materials. Usually, rotatory screens are used. The primary method covers shredding and sieving while in the secondary method crushing and pulverization with dry air cyclone with dust collectors are used. Gravity separation: The materials are separated on the basis of density. Denser the material it will leave itself on the bottom and *vice versa*. But it also depends on the size as well. Just not the force of gravity help separates the material but the viscous liquid tetrabromoethane; the heavy liquid separates the plastics and ceramics from the metals. Heavy media separators, water, air flows along with the gravity separators are used to separate the e-waste. The air blowers separate the materials by passing of the air and separating on the basis of difference in their densities. Magnetic separators: The low-density drums separate the magnetic (ferrous metals) from the non-magnetic (non-ferrous metals) materials. The main disadvantage is that the aggregation of the particles and to some extent the contamination of Cu in ferrous metal separation. In the separation of PCBs two phases of magnetic fields are used; in the first stage 700G low magnetic field is applied which separates 83% Ni and Fe in the magnetic fraction and 92% Cu in the non-magnetic fraction. The second stage applied 3000G of magnetic energy and separated more concentrates of Cu than Ni and Fe. Electrostatic separation: Different electrical conductivity or resistivity for materials separates. It depends on the polarity and charge acquired by the materials. Corona is successful in separating on the basis of polarity. Tribo-electricity separates with similar conductivities. Eddy current separates by the phenomena of falling of the particles by the combined forces of gravitational, centrifugal, frictional and magnetic. The magnetic force greatly deflects iron in the mixture of waste. It removes Al, Cu, broken glasses and non-ferrous metals. Eco-friendly, no harmful emissions with ability to separate both metallic and non-metallic is corona electrostatic method. The metallic ones get attached to the electrode while non-metallic ones get charged and fall directly in the storage bins. 0.6-1.2mm size is suitable for the removal by electrostatic separator. Froth floatation: It deals with the hydrophobicity of the material. Poor recovery of precious metals of 74μm fine fraction is retained. Collector less floatation with gasoline is best for the separation of non-metals mainly polymers. The main advantage of the physical recycling is that makes the process simple, eco-friendly, investment and capital cost low with diversification of the methods. Dust generation, metal loss, hazardous waste leaching are some challenges in physical recycling which have to be fore looked.

Chemical Recycling- Polymer waste is broken down into monomers by chemical application. Pyrolysis, gasification, supercritical liquids, and hydrogenolytic degradation. It separates the organic and the metallic materials. High purity metals obtained by high temperature (400-900°C) heating in absence of oxygen *i.e.,* pyrolysis. The organic materials are polymerised and decomposed into low molecular weight such as oil, tars *etc* while the inorganics or the metals remain as such. The product obtained is char and low value components. Further the product is sent for further processing to be used in metallurgy. The major short coming is the presence of dioxins which is present as a precursor in the pyrolysis oil and can be reduced by the addition of Fe_2O_3 and $CaCO_3$. The epoxy layers are removed by the supercritical fluids.

Centrifugal separation and vacuum pyrolysis are proven to be the best for separating solder and organic materials from PCBs (250-370°C) due to short vapor time in the reactor and lower decomposition temperature which reduces the occurrence and intensity of secondary reactions. The recovery of solder leads to the direct use of it again and recovery of Sn and Pb by refining can be carried out. Combustion recycles the glass into glass slay and reduces the efficiency of the process. The main purpose of chemical recycling is to convert materials into fuel. The chemical recycling has an advantage over physical recycling that the BFR's are transformed into monomers and the heavy metals are removed as a residue. The chemical recycling helps to reduce all the hazardous and toxic component out of the waste (Kaya 2016).

PET bottle: Physical/ Mechanical Recycling- From collection to sorting, washing and crushing/grinding of the waste. Let us now understand taking PET bottles which are most commonly found abundantly in waste plastics. Collection methods employed are drop-off, curbside, refill and deposit system. The main observation during collection is the quantity and extent of contamination. After sorting they are compressed in to bales where they are transported easily. They are processed further for final product. The further processing comprises of: -a) Solid state post condensation- Transitional heating of PET in between glass and melting temperature of reactor. Condensation reaction is carried out in amorphous state and in the terminal group of the terminal group at 200-240°C under vacuum conditions which removes the by-products. b) Chain extenders: Low molar mass compound with different functional groups are added to the terminal of the polymer chain to increase the length and the molecular weight. Most widely used epoxy based-oligomer with multifunctional purpose is joncryl (Baden Aniline and Soda Factory). Chemical recycling: it is all about transforming the waste (feedstock) into a used product mainly fuels or chemicals. The feedstock has sufficient high hydrocarbon content and higher carbon efficiency with no oxygen and less moisture. Pyrolysis, gasification, fluid-catalysed cracking and hydrocracking.

Chemical recycling- PET it transforms the polymer form of PET to monomer form such as terephthalic acid (TPA), dimethyl terephthalate (DMT), bis(hydroxyethylene)terephthalate (BHET) and ethylene glycol (EG). Depolymerization occurs to transform into oligomers. Methanolysis, glycolysis, hydrolysis and ammonolysis *etc* are few reactions of depolymerization. In methanolysis: the treatment of PET in methanol at high temperature (180-280°C) and pressure (20-40atm) forming products EG and DMT. These products can be used as plasticizers, crosslinkers, chain extenders, corrosion inhibitors and precursors. Hydrolysis: breaking down the polymer bond in water in neutral, acidic, and basic conditions at high temperature forming TPA and EG as products. Slow reaction due to nucleophile impairment and impure TPA in the product. Glycolysis: one of the most widely used commercial methods for PET recycling. α, ω- dihydroxy materials (polyols) are formed. Their application is in epoxy resins, vinyl esters, polyurethanes, *etc*. Transesterification of PET in presence of excess of glycol at 180-250°C BHET as an end product (Ragaert *et al*. 2017). Glycolysis and efficiency improvement is carried out using various catalysts such as ionic liquids (1-butyl-3-methylmidazolium bromide), metal salts (sodium carbonate, sodium bicarbonate, sodium sulfate, and potassium sulfate), hydrotalcites, and enzymes (*Fusarium solani pisi* degradation by cutinase and *Candida antarctica* lipase) (Khoonkari *et al*., 2015).

3.6. Waste Composition After Recycling

The mixed waste coming in to the facility has portions of recyclable and non-recyclable materials. The waste is supposed to increase by 2025 by 2.2 billion tonnes (acc. to world bank). USEPA data 2015 mentions; the total waste generated only 25.8% is recycled, 8.9% is composted, 52.5% sent to landfill and 12.8% is sent for recovery and combustion. The incoming waste has some materials as contaminants which are removed by the physical and chemical recycling processes for various materials such as plastics, glass, e-waste, metals, *etc*. Further the recyclable materials are used for various other processes which turn out be more valuable products than the virgin raw materials. The analysis of waste composition after recycling helps portraying clear picture and uniformity in the data for complete and meaningful information (Shaw *et al*. 2006).

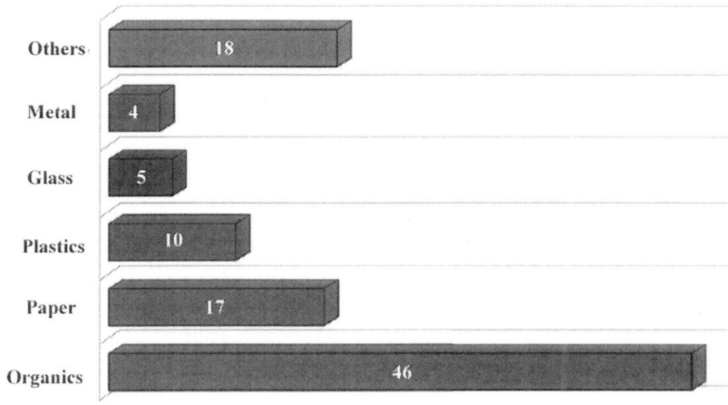

Fig. 7: Global trends of waste generation (Source: USEPA DATA)

The above figure depicts the trend of the waste managed globally. The composition of waste after recycling can be explained by citing an example of recycling plastic waste. PYROPLEQ processing technology adopted in Austria, Germany, Korea, Italy and Switzerland. The temperature for pyrolysis is 450-500°C with heated rotary drums with gas combustion at 1200°C. Akzo process adapted by Netherlands with circulating fluidised bed reactor system with subsequent combustion. In the waste mixture majority of PVC is treated with HCl, CO, H2, CH4 as the by- products with hydrocarbons and fly ash production. Other technology is ConTherm and automotive shredder residue (ASR) at 500-550°C. The residues are screened and sorted for the recovery of many metals. NRC technology is employed for the extraction of metals. The aim is to produce purified CaCl2 rather than HCl. The mixed waste such as cables, flooring, *etc* rich in PVC are only fed and CaCl2, coke, fuels and heavy metals for metal recycling are produced. Pyrolysis (chemical recycling) it offers many advantages such as the production of fuels from the feedstock, no flue gas for clean-ups, alternate to landfill and reduces emissions of GHG's and CO2. The fuel produced are of high calorific value and hence marketed as a profitable product to produce heat and electricity. It is one of the viable sources of recycling (Al-Salem *et al.* 2009). Following table.4. (a) and (b) explains the composition of plastic waste after recycling undergoing chemical recycling process of pyrolysis and cracking respectively.

Table.4. (a) Waste composition after recycling (by pyrolysis)

Materials	Components after recycling (pyrolysis) (wt.%)				References
Plastics	Oil	Gas	Solid residue	Others	(BREMS *et al.* 2012)
PS	24.6	9.9	0.6	64.9 styrene	
PE	50.3	7.6	0.1	42 waxes	
Polyester	40	50.8	7.1	2.1 water	
PVC	28.1	6.8	8.8	56.3 HCl	
PMMA	1.4	1.25	0.15C	97.2MMA	

Table.4. (b) Waste composition after recycling (cracking process)

Materials	Components after recycling (chemical or cracking process)								References (Al-Salem *et al.* 2009)
	Polyolefins	PS	PET	PVC	Total plastic content	Ash	Moisture	Metal Pieces	
Plastics	80	15	3	2	92	2	0.5	0	

3.7. Cost Analysis for Recycling

3.7.1 Concept of Cost Analysis: Aim and Need

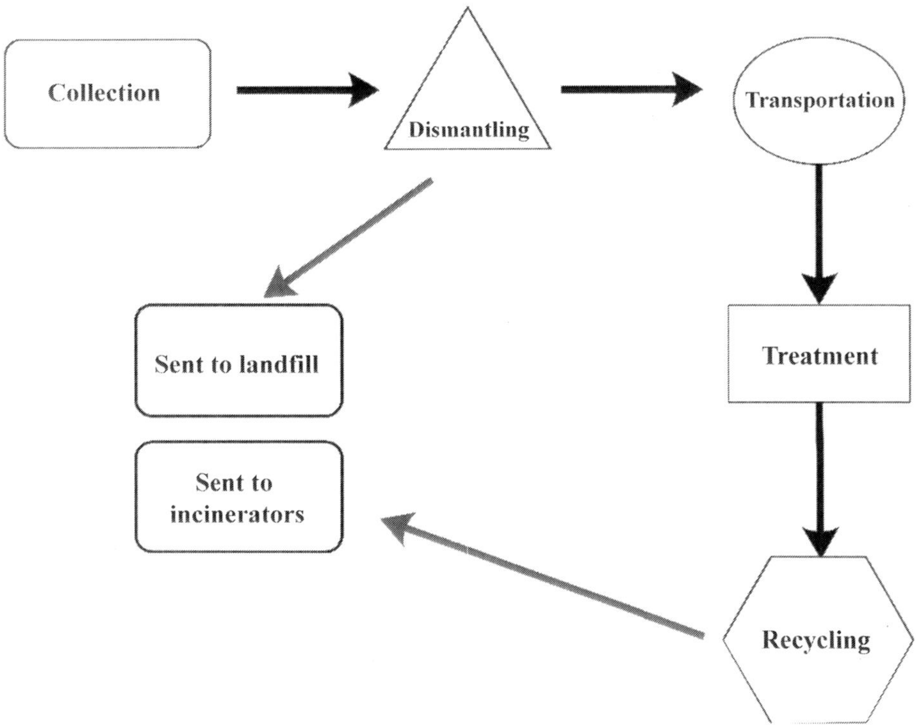

Fig. 8: Recycling process with step-wise cost analysis (Source: Farel *et al.* 2013)

Cost Benefit Analysis (CBA) is a formal way of analysing the benefits and the costs incurred by the companies. It helps in comprehensive assessment of feasibility study in terms of environment, social, economic and technical aspects. In 1902 for the first time the method was implemented for the economic feasibility of irrigation project in USA this method was employed. The main objective is to analyse the cost to evaluate its performance, estimate the cash flow process, determine the gap and help in identifying more attractive options, *etc*. The whole process must be followed by estimating the total operational commitment which includes the initial capital cost or investments then the material flow, its maintenance and operation. Identifying the funding resource and financial sustainability is also one of the important aspects while analysing the cost. The estimation must be done close to the economic lifetime and taking in account the present and the long - term impacts (Callan and Thomas 2001). Let us now understand the concept of cost recycling by taking an example of End-of-life vehicles (ELV) with its glass recycling. In Europe around 8 million vehicles per annum are discarded as a waste. The glass from the discarded vehicle which is 2.8% of the weight are also discarded. The glass of windshield, windows and back windows are all regarded as glass glazing. Usually 90-95% of the glass glazing is sent to the landfill. Sending the ELV glass to the landfill is not the issue. It is further mixed with other materials and residues. The treatment and recycling of glass is non-feasible and expensive deal. The best option available is collecting the ELV glass and transport it to the glass recycling facility where it is processed and treated to purified so as to use the glass as a new fresh raw material know as cullet. Cullet is eco-friendly and economic material due to its less energy consumption in furnace and cheaper in price than the raw material respectively. Lots of natural resources are conserved due to the glass cullet formation. CBA helps in determining the policy proposal and promotes the economic efficiency and measures the both positive and negatives in terms of benefits and cost respectively. Access if the costs exceed the benefits and vice-versa. CBA is helpful in analysing the financial status and the material recycling flow. It helps in waste chain management. Cost benefit analysis is useful because the decision-making process depends on both the pros and cons of the process, it maintains internal consistency within the internal and external flow of costs (Farel *et al*. 2013).

3.7.2 Calculating the Recycling Cost

The recycling cost is affected by various factors such as material weight, material recycling, cost of storage and disposal, *etc*. The total recycling cost is calculated as follows:

The cost to be included for estimation is the cost of disassembly that is sorting, treating *etc.* and cost of disposal and recycled materials as products. The inference we draw regarding recycling cost is the profit gained by selling the recyclable goods - the total cost incurred during storage, transportation and disposal. Recycling Cost = $\sum$ Revenue - $\sum$Cost.

$\sum$Recycling cost = $\sum$Cost of Recyclable products - ($\sum$Cost of treatment + $\sum$Cost of waste + $\sum$Cost of disassembly).

Evaluation of cost of Recyclable products = weight (kg) × Price (formula same for Cost of treatment and Cost of waste).

Evaluation of Cost of disassembly = disassembly × time (rate × h).

Longer the time for disassembling; more costly will be the recycling cost of recyclable and reusable materials. The final recycling cost depends on the recyclable products formed and divided and used differently. The grade of the product affects the price of the product. In case if the low-grade materials such as plastics, steel, *etc* are unable to separate for the high grade then the price will automatically be low and hence combined with unwanted or hazardous waste will be counted as one. Further removing the useful material out of the unwanted or hazardous waste material the additional treatment cost will be added (Karwasz *et al.* 2016).

The main target of recycling is to increase the market value of the recyclable products almost equal or more than the virgin materials as such and reduce the pressure and volume and use of virgin materials. Competitive advantage to the recyclable products in terms of quality and its life (International 2001).

We will try understanding the Cost benefit analysis of Paper recycling- Benefits of Paper recycling

1. The value of the product: There is considerable increase in the value of the product specifically the secondary fibre material. The indirect tax on the paper by the government has also helped the value price of recycled paper.

2. Avoidance of the additional cost of disposal: The recycling of paper led to the reduced burden on the other disposal options. Landfilling and incineration options can create environmental and health impacts. The cost due to them are avoided and are almost negligible. Value of the land to be acquired for landfill then the technologies for the incineration and the maintenance for the release of gas and production of unwanted products and by-products at high temperature combustion. These issues and costs are all dealt by the recycling method. Though the recycled

paper has not been able to replace the use of the virgin raw materials for the production of paper as there is this limitation of recycling the paper not many times due to the length of the fibres.

3. Reduced collection cost: Recycling more than a ton waste automatically reduces the collection cost by reducing the total amount of waste to be collected. This leads to a reduction in truck trips to collect waste and send it to recycling centres and thus the avoided cost must be deducted from the cost of collected recyclables.

The costs associated with the recycling are:

1. Collection and Recovery cost- The collection cost includes the expenditure/cost of the resources. The cost borne by the operator is variable and affects the overall cost. The collection and recovery cost refer to the cost difference due to the waste brought by the person to the recycling centre or the recycling centre employ someone to carry and collect the waste to the recycling centre. Avoided collection cost from regular garbage collection are the positives and the collection cost for recycling are the negatives and an add on to the cost of recycling.
2. Environmental cost: These will deeply emphasise and effect the external costs with different phases of recycling process. The environmental cost refers to the transportation cost. The waste transported to the landfill and the pollution cost association that is the pollution caused during bleaching, deinking process *etc* must be mentioned as addition to the cost and the operational cost associated with the recyclable products to bring them to the virgin raw materials is not regarded as the cost but the benefit and this situation is controversial. Usually, the operational cost is the goods/materials sold + expenses. If the goods sold are at low price then it will be included as an additional cost in the recycling. Hence some factors/ cost can be both beneficial and disadvantage and all depends on the market price the recyclable products are sold. As shown in the factors affecting the cost of recycling; The market value of a product is one of the most important factors in determining recycling cost analysis (Hanley and Slark 1994).
3. Analysing operation and transportation cost by the example of glass cullet. The cost is divided into different processes such as dismantling, collection, storage, transportation and treatment. These costs are dependable on different parameters. The collection and transportation costs are the ones effecting the recycling cost process. The reduction of collection and cost can be done by sufficient dismantling, storage,

treatment units with increase in the volume reduction is beneficial for collection and transportation cost analysis (Farel *et al.* 2013).

Let us take the example of electrical and electronic waste estimation (PĂCESILĂ *et al.* 2015). The recycling cost is affected by various factors such as material weight, material recycling, cost of storage and disposal, *etc.* Recycling cost evaluation by Carroll in 1992 studied that recycling cost per household is higher when services provided by the municipalities rather than the private company. In Folz (1995) paper; the relation between public participation and decrease of average cost of both curbside and drop off (types of collection systems). The most significant conclusion drawn by folz in 1996 was that; higher the recycling amount lead to the decrease in the change in unit cost and decline in unit cost (which explains the scale economics). Even estimated the cost of recycling for the towns with less than 5000 population which is less in comparison to collection and disposal cost hence it explains the importance of recycling waste economically (Callan and Thomas 2001). The major restriction to the analysis is that we often try including the cost incurred by various stages of recycling up to the final recyclable product further in-depth analysis for the other disposal options like landfilling and incineration must too be included.

3.8 Public Participation

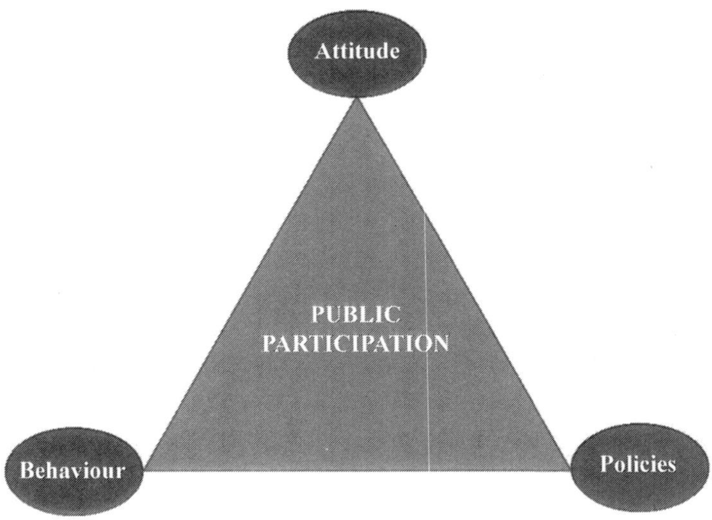

Fig. 9: Public Participation Triangle: Factors determining the mass participation (Source: Timlett and Williams 2008)

Participation is a free process and a part taking in forming and taking an action. It is a process where the communities, stakeholders and policy makers come together and implement, take further actions to bring a specific change. The participation acts as a catalyst to success of an undertaking in a community (Kaloki 2015). The attitude, behaviour and policy implementation are the main determining factors for the mass participation. Public play a significant role in success of the waste management policies. D.J. Lober (1996) also states the attitude and behaviour of the people/ public which likely effects the recycling. Intrinsic attitude *i.e.,* internal feel to manage the waste and carry out recycling is really important and an add on to the initiative in proceeding with the recycling. Similar factors were highlighted (Timlett and Williams 2008) about providing incentives and providing the convenient facilities of collection and transportation to the public/ community. D.J. Lober (1996) also highlighted that the socio-economic and demographic variables are not the influencing factors in public recycling activities (Lober 1996). The public participation is a great influence on the recycling practice. The case study in China-Beijing for recycling e-waste stated that the low-income group people rarely participated in the recycling. This is due to the poor transportation facilities and the low-income group people preferred more reusing the product than the high income and using more of the second-hand products. Attitude and time are the variable factors in case of recycling. Senior citizens were involved in most of the recycling programmes than the younger generation. Educating consumers on recycling can enhance the public participation. Recycling habits, recycling facilities and services and economic benefits are the contributing factors in public participation in recycling as identified by the study. Improper management and implementation of recycling of waste also affect the public role in recycling (Wang *et al.* 2011). Attitude Recycling: Self-participation in the recycling process. Exaggerated behaviour of the people was observed *i.e.,* what they say and what they do. It is important that the positive attitude turns to be an effective thought in application for recycling every waste item. It has become evident to apply practically the recycling practice which is now being recognised by the policy makers. Attitude is not the only influencing factor which effects the recycling. The other important point is the behavioural recycling. Behavioural recycling: It is important to strategize and form policies which can bring change in the ingrained habits. It is controlled by certain factors such as:

1. Providing infrastructure, removal of barriers, education,
2. Introducing taxes, penalties, rewards and incentives,
3. By communication and feedback to keep a check in policies and
4. Setting an example (Timlett and Williams 2008).

Provision and access to education, infrastructure is an essentiality in public participation. As discussed above the willingness in people is important. With respect to the above points education and awareness among the masses about the need, aim and benefits of recycling is crucial. Incentives, rewards and at the same time penalties against unfollowing the regulations. Communicating and giving feedback responses helps the authorities and the policy makers to improvise and bring out the best and effective schemes and programmes. Setting an example by the community or group of people helps to realise the effectiveness in the strategies amended by the government. Case study of the Labuan where public participation played an important role in recycling. The Labuan municipal corporation promoted and encouraged their citizens by involving them in recycling programmes, giving them the incentives. The participation helped the citizens to inculcate the recycling practices at their homes and make it a daily routine (Fatma *et al.* 2016). Similar case studies and successful stories highlighting the key conclusions for further improvement have been mentioned in various other papers. Most widely the recycling of plastics, households and e-waste have been investigated (Akil *et al.* 2015), (Lober 1996), (Wang *et al.* 2011), (Mcdonald and Ball 1998). The inference drawn is that public plays an imminent role in recycling process of any waste. It the intrinsic approach *i.e.,* self-enthusiasm and add-on is the extrinsic approach *i.e.,* providing the infrastructures, facilities, power, *etc.*

References

Akil, A.M., Johar, F., Siong, H.C., 2015. Household Participation In Recycling Programs: A Case Study from Malaysia. Tata Loka. 17(2), 64-75. https://doi.org/10.14710/tataloka.17.2.64-75 Mcdonald, S., Ball, R., 1998. Public participation in plastics recycling schemes. Resources, Conservation and Recycling. 22 (3-4), 123-141. https://doi.org/10.1016/S0921-3449 (97) 00044-X

Allesch, A., Brunner, P.H., 2014. Assessment methods for solid waste management: A literature review. Waste Management & Research. 1-13. https://doi.org/10.1177/0734242X14535653

Bharti, V., Vikrant, K., Goswami, M., Tiwari, H., Sonwani, R.K., Lee, J., Tsang, D.C.W., Kim, K.H., Saeed, M., Kumar, S., Rai, B.N., Giri, B.S., Singh, R.S., 2019. Biodegradation of methylene blue dye in a batch and continuous mode using biochar as packing media. Environ. Res. 171, 356–364. https://doi.org/10.1016/j.envres.2019.01.051

Bom, U.B., Belbase, S., Lila, R.B., 2017. Public Perceptions and Practices of Solid Waste Recycling in the City of Laramie in Wyoming, U.S.A. Recycling. 2,11. https://doi.org/10.3390/recycling2030011

Borodin, Yu V., Aliferova, T.E., Ncube, A., 2015. Waste management through life cycle assessment of products. Materials Science and Engineering. 81. https://doi.org/10.1088/1757-899X/81/1/012085

Brems, A., Baeyens, Jan., Dewil, R., 2012. Recycling And Recovery of Post-Consumer Plastic Solid Waste In A European Context. Thermal Science. 16(3), 669-685. https://doi.org/10.2298/TSCI120111121B

Callan, S.J., Thomas, J.M., 2001. Economies of Scale and Scope: A Cost Analysis of Municipal Solid Waste Services. Land Economics. 77(4), 548-560. https://doi.org/ https://doi.org/10.2307/3146940

Coelho, L.M.G., Lange, L.C., Coelho, H.M.G., 2016. Multi-criteria decision making to support waste management: A critical review of current practices and methods. Waste Management & Research. 1-26. https://doi.org/10.1177/0734242X16664024

Daum, K., Stoler, J., Grant, R.J., 2006. Toward a More Sustainable Trajectory for E-Waste Policy: A Review of a Decade of E-Waste Research in Accra , Ghana. International Journal of Environmental Research and Public Health. 14, 135. https;//doi.org/10.3390/ijerph14020135

Dhote, L., Mandpe, A., Paliya, S., Kumar, S., Pandey, R.A., Kumar, R., 2020. Characterization of distillery sludge for its thermal properties and ascertaining its utilization as a low-cost fuel. J. Clean. Prod. doi: 10.1016/j.jclepro.2020.120872

Drzyzga, O., Prieto, A., 2018. Plastic waste management, a matter for the 'community.' Microbial technology. 0(0), 1-3. https://doi.org/10.1111/1751-7915.13328

European Commission, 2004. Methodology for the analysis of Solid waste. https://www.wien.gv.at/meu/fdb/pdf/swa-tool-759-ma48.pdf (accessed on 18 June 2019)

Farel, R., Yannou, B., Ghaffari, A., Leroy, Y., 2013. A cost and benefit analysis of future end-of-life vehicle glazing recycling in France: A systematic approach. Resources, Conservation and Recycling. 74, 54-65. https://doi.org/10.1016/j.resconrec.2013.02.013

Fatma, S.A., Latifah, A.M., Mariani, H.N.O.A., SJHA. Public Participation Recycling Through Econ Benefit Case Study Fed Territ Labuan. 2016.

Finnveden, G., Björklund, A., Moberg, A., Ekvall, T., 2007. Environmental and economic assessment methods for waste management decision-support: possibilities and limitations. Waste Management & Research. 25, 263-269. https://doi.org/10.1177/0734242X07079156

Gronostajski, J., Matuszak, A., 1999. The recycling of metals by plastic deformation: an example of recycling of aluminium and its alloys chips. Journal of Material Processing Technology. 93, 35-41. https://doi.org/10.1016/S0924-0136(99)00166-1

Hanley, N., Slark, R., 1994. Cost-benefit analysis of paper recycling: a case study and some general principles. Environmental Planning and Management. 37(2), 187-197. https://doi.org/10.1080/09640569408711969

Haupt, M., Kägi, T., Hellweg, S., 2018. Life cycle inventories of waste management processes. Data in Brief. 19, 1441-1457. https://doi.org/10.1016/j.dib.2018.05.067

Hoornweg, D., Bhada-Tata, P., 2012. WHAT A WASTE A Global Review of Solid Waste Management, The World Bank. https://openknowledge.worldbank.org/handle/10986/17388 Puna, J.F., Santos, M.T., 2010. Thermal Conversion Technologies for Solid Wastes: A New Way to Produce Sustainable Energy. https://doi.org/10.5772/8461

Imteaz, M.A., Ali, M.M.Y., Arulrajah, A., 2015. Possible environmental impacts of recycled glass used as a pavement base material. Waste management and Research. 30(9), 917-921. https;//doi.org/10.1177/0734242X12448512

International, RDC-Environment and Pira. "Evaluation of costs and benefits for the achievement of reuse and recycling targets for the different packaging materials in the frame of the packaging and packaging waste directive 94/62/EC" – proposed draft final report, RDC-Environment & Pira International. May 2013. https://ec.europa.eu/environment/waste/studies/packaging/public_discussion.pdf

Jim´enez, L.F., Dom´ınguez, J.A., Vega-azamar, R.E., 2018. Carbon Footprint of Recycled Aggregate Concrete. Advances in Civil Engineering. 0-6. https://doi. org/10.1155/2018/7949741

Kaloki, F.K., 2015. Role of public participation in solid waste management in Mlolongo town. Erepository.uonbi.ac.ke

Kaplan, R.S., Ness, H., 1987. Recycling of Metals. Conservation and Recycling. 10(1), 1-13. https://doi.org/10.1016/0361-3658(87)90002-6

Karwasz, A., Dostatni, E., Diakun, J., Grajewski, D., Wichniarek, R., Stachura, M., 2016. Estimating The Cost of Product Recycling with the use of Ecodesign Support System. Management and Production Engineering Review. 7(1), 33-39. https://doi.org/10.1515/mper-2016-0004

Kaya M. Recovery of metals and nonmetals from electronic waste by physical and chemical recycling processes. Waste Management. 57, 64-90. https://doi.org/10.1016/j.wasman.2016.08.004

Kennedy, S.M., Copes, R., Bartlett, K.H., Brauer, M., 2004. Point-of-sale glass bottle recycling: indoor airborne exposures and symptoms among employees. Occupational Environmental Medicine. 61, 628-635. https://doi.org/10.1136/oem.2003.009753

Khoonkari, M., Haghighi, A.H., Sefidbakht, Y., Shekoohi, K., Ghaderian, A., 2015. Chemical Recycling of PET Wastes with Different Catalysts. International Journal of polymer science. 11. https://doi.org/10.1155/2015/124524

Kiker, G.A., Bridges, T.S., Varghese, A., Seager, T.P., Linkov, I., 2005. Application of Multicriteria Decision Analysis in Environmental Decision Making. Integrated Environmental Assessment and Management. 1(2), 95-108. https://doi.org/10.1897/IEAM_2004a-015.1

Kumar, A., Holuszko, M., Espinosa, D.C.R., 2017. E-waste: An overview on generation, collection, legislation and recycling practices. Resources, Conservation and Recycling. 122, 32-42. https://doi.org/10.1016/j.resconrec.2017.01.018

Kumar, S., Bhattacharyya, J.K., Vaidya, A.N., Chakrabarti, T., Devotta, S., Akolkar, A.B., 2009. Assessment of the status of municipal solid waste management in metro cities, state capitals, class I cities, and class II towns in India: An insight. Waste Manag. 29, 883–895. https://doi.org/10.1016/j.wasman.2008.04.011

Ling, T., Poon, C., Wong, H., 2013. Management and recycling of waste glass in concrete products: Current situations in Hong Kong. Resources, Conservation and Recycling. 70, 25-31. https://doi.org/10.1016/j.resconrec.2012.10.006

Lober, D.J., 1996. Municipal Solid Waste Policy and Public Participation in Household Source Reduction. Waste management and Research. 14, 125-143. https://doi.org/10.1177/0734242X9601400203

Mandpe, A., Kumar, S., Lakshmikanthan, P., Hettiarachchi, H., 2019a. Mining for recovery as an option for dumpsite rehabilitation: Case study from Nagpur, India. J. Environ. Eng. Sci. 15, 52–60. doi:10.1680/jenes.19.00021

Maschio, S., Tonello, G., Furlani, E., 2013. Recycling Glass Cullet from Waste CRTs for the Production of High Strength Mortars. Journal of Waste Management. 0-8. https://doi.org/10.1155/2013/102519

Methodology for the analysis of Solid waste, EU, 2004. https://www.wien.gv.at/meu/fdb/pdf/swa-tool-759-ma48.pdf

Ministry of Urban Development, 2016. MUNICIPAL SOLID WASTE MANAGEMENT MANUAL. http://www.swachhbharaturban.gov.in/ (accessed on 20 June 2019)

Mirkovic, I.B., Majnaric, I., Bolanca, Z., 2015. Ecological Sustainability and Waste Paper Recycling. Procedia Engineering. 100, 177-186. https://org.doi/10.1016/j. proeng.2015.01.356

Muttamara, S., Visvanathan, C., Alwis, k.u., 1994. SOLID WASTE RECYCLING AND REUSE IN BANGKOK. Waste management and Research. 12, 151-163. https://doi.org/10.1177%2F0734242X940120020

Owen, K.C., 1998. A Pricing Strategy for a Recycling Industry. Corporate Environmental Strategy. 5(2), 42-50. https://doi.org/10.1016/S1066-7938(00)80065-2

Păcesilă, M., Ciocoiu, C.N., Colesca, S.E., Burcea, Ş.G., 2015. An Overview of Cost Benefit Analysis for Weee Recycling Projects. "Management and Innovation For Competitive Advantage". 321-332. https://conference.management.ase.ro/archives/2015/pdf/36.pdf

Ragaert, K., Delva, L., Geem, K. V., 2017. Mechanical and chemical recycling of solid plastic waste. Waste Management. https://doi.org/10.1016/j.wasman.2017.07.044

Shaw, P.J., Lyas, J.K., Hudson, M.D., 2006. Quantitative analysis of recyclable materials composition: Tools to support decision making in kerbside recycling. Resources Conservation and Recycling. 48, 263-279. https;//doi.org/10.1016/j.resconrec.2005.12.013

Al-Salem, S.M., Lettieri, P., Baeyens, J., 2009. Recycling and recovery routes of plastic solid waste (PSW): A review. Waste Management. 29(10), 2625-2643. https://doi.org/10.1016/j.

Timlett, R.E., Williams, I.D., 2008. Public participation and recycling performance in England: A comparison of tools for behaviour change. Resources Conservation and Recycling. 52, 622-634. https://doi.org/10.1016/j.resconrec.2007.08.003

UN Habitat for Better Urban Future, 2012. Recycling and disposal of municipal solid waste in low and middle-income countries: Perspectives for municipal managers and environment agencies. https://mirror.unhabitat.org/downloads/docs/Recycling.pdf (accessed on 12 June 2019).

Wainaina, S., Awasthi, M.K., Sarsaiya, S., Chen, H., Singh, E., Kumar, A., Ravindran, B., Awasthi, S.K., Liu, T., Duan, Y., Kumar, S., Zhang, Z., Taherzadeh, M.J., 2020. Resource recovery and circular economy from organic solid waste using aerobic and anaerobic digestion technologies. Bioresour. Technol. 301, 122778. https://doi.org/10.1016/j.biortech.2020.122778

Wang, Z., Zhang, B., Yin, J., Zhang, X., 2011. Willingness and behavior towards e-waste recycling for residents in Beijing city, China. Journal of Cleaner Production. 19(9-10), 977-984. https://doi.org/10.1016/j.jclepro.2010.09.016

Weng, Y.C., Fujiwara, T., 2011. Examining the effectiveness of municipal solid waste management systems: An integrated cost – benefit analysis perspective with a financial cost modeling in Taiwan. Waste Management. 31(6), 1393-1406. https://doi.org/10.1016/j.wasman.2011.01.016

Wu, B., Zhou, M.H., 2009. Recycling of waste tyre rubber into oil absorbent. Waste Management. 29(1), 355-359. https://doi.org/10.1016/j.wasman.2008.03.002

Zhang, X., Li, H., Cao, Q., Jin, L., Wang, F., 2018. Upgrading pyrolytic residue from waste tires to commercial carbon black. Waste management and Research. 79, 1-9. https;//doi.org/10.1177/0734242X18764292

4

Legislations and Management Practices of Municipal Solid Waste in Developing Countries

***Ashootosh Mandpe*[a,b,c*] *and Lakshay Tyagi*[b,c]**

[a]*Department of Civil Engineering, Indian Institute of Technology Indore Indore 453 552, India*
[b]*Academy of Scientific and Innovative Research (AcSIR) Ghaziabad- 2010 02, India*
[c]*CSIR-National Environmental Engineering Research Institute (CSIR-NEERI) Nehru Marg, Nagpur 4400 20, India*

ABSTRACT

Solid Waste Management is a challenge for the municipal authorities in developing countries mainly due to lack of policy planning and enforcement and overstressed technological, infrastructural, and human resources for the handling of behemoth quantum of solid waste generated. For the Indian scenario, a comprehensive literature assessment of developments in pro-environmental and waste management policies and legislation from the 1970s to 2020 was conducted. Also, a brief insight into the waste management practices followed in the major metropolitan cities of the three developing nations; Great Buenos Aries (Argentina), Mexico City (Mexico), and Manila (Indonesia) were made. An integrated solid waste management strategy for Indian conditions was drawn out in complete conformity with existing waste handling and management regulations. The government of India (GoI) made a strong paradigm shift in the enacted waste management legislation in the year 2016 by inducting long-awaited recommendations into the jurisdiction. Major inductions in the enacted waste management legislations along with key governmental initiatives in the realm of waste management such as Swacch Bharat Mission (SBM) were reported. This book chapter addresses the chronological development in the environmental and waste management legislation of India and prescribes the fundamental duties/role required to be played by the Urban local bodies (ULBs), Gram panchayats (GPs), and society to achieve environmental sustainability and clean surroundings.

Solid Waste Management is a challenge for the municipal authorities in developing countries mainly due to lack of policy planning and enforcement and overstressed technological, infrastructural, and human resources for the handling of behemoth quantum of solid waste generated.

Keywords: Anthropogenic Waste Legislations, Integrated Solid Waste Management, Swachh Bharat Mission, Solid Waste Literacy

4.1. Introduction

Presently, in the world, there are 195 countries according to the United Nations, and the number would even go up to 247 if we consider the disputed territories like Isle of Man, Hong Kong, etc. (United Nations, 2016) (Kershner, 2020). But the majority of the countries are still struggling to provide basic life-supporting amenities like nutrition, shelter, and employment. Conjointly, termed as emerging and developing countries such as India, Vietnam, Venezuela, Utopia, etc. (Ahmed and Shimada, 2019). Most of the developing countries are situated mainly in the continent of Africa, Asia and South America. They have got independence for the European countries like England, France, Portugal and Spain in the mid and late 20th century.

The course of development of a country in the pursuit of upliftment of its citizens is laid on a tripod of industrialization, urbanization and mechanized agriculture. Although these activities fulfil the basic needs of the masses like employment, shelter, and nutrition at the same time, these activities are the major contributors of the Municipal Solid Waste (MSW) (Kumar *et al.*, 2017) transport, treatment and disposal. Current systems in India cannot cope with the volumes of waste generated by an increasing urban population, and this impacts on the environment and public health. The challenges and barriers are significant, but so are the opportunities. This paper reports on an international seminar on 'Sustainable solid waste management for cities: opportunities in South Asian Association for Regional Cooperation (SAARC. And it has been observed that countries where the rate of industrialization and urbanization is rapid the problem of MSW is more acute as compared the countries with lower industrialization and urbanization rate.

In India, waste management system founds its roots deep into its history as, during the ancient era, the Vedas, the Upanishads, the Smritis and all other holy texts were devised which worships 'nature' as a God, even the planets including earth and the stars of the solar system were adored by the masses since times immemorial. Also, the physical traces of a sophisticated water-flushed sewage system was discovered in the mounds of Harappan civilization that used to flourish way back 4,000 years ago spanning between 2600 to 1900 BCE (Antoniou *et al.*, 2016). However, in modern time, the state of solid waste management is kept ignored. As even after the independence, the newly formed Government of India (GoI) declared sanitation and water supply as a goal in the first five-year plan (1951-56), though nothing significant could be achieved on the ground and after the nearly 33 years of lackadaisical approach the GoI had come up with a country's first nationwide sanitation programme in the year 1986 under the name of Central Rural Sanitation Programme (CRSP), this attempt of GoI was gone into the vein as it was a supply-driven, highly subsidized and infrastructure oriented programme. Again, GoI had bounced back and came up with Total Sanitation Campaign (TSC) in the year 1999 intending to improve the general quality of life and to achieve total sanitation coverage by the year 2012 (Ganguly, 2008). However, falling short of the targets it was renamed and relaunched in the year 2012 as Nirmal Bharat Abhiyaan. In the year 2014, the GoI had launched Swachh Bharat Mission (SBM) at the foremost plan for implementation of sanitation that focuses on inducing behavioural change among the citizens regarding waste management, cleanliness and open defecation in all rural and urban dwellings of the country with two variants, i.e. SBM-Gramin and SBM-Urban respectively (Mohapatra, 2019).

Apart from the sanitation, the GoI has revealed its intention for a complete overhaul in the legislation concerning the MSWM and establishment of the requisite infrastructure for enforcement of these rules. However, the apart from the presence of a robust set of legislations and institutions the state of MSWM is still disappoints in the majority of the cities/towns and rural dwelling mainly due to the lackscadial approach of the Urban Local Bodies (ULB's) / Gram Panchayats (GP's) in the proper enforcement of the laid rules and inducing the behavioural change in the masses regarding the MSWM.

4.2. Specific Targets for Waste Management in India

As far as developed countries are concerned in the formulation and planning of waste management strategy, major emphasis is given on the protection of the environment and to achieve sustainability without compromising with the standard of living of the citizens. All other goals are, by far, less critical. On the other hand, for developing countries, especially in the case of India, the situation is quite complex in the field of waste management. Though India has a legacy of reverencing nature and possesses a history of launching of various nationwide sanitation programmes time to time since its independence, however key issues under waste management remained unanswered and needs to be acknowledged through robust policies and on-ground implementation.

The movement of urbanization in India had gained momentum in the past decades as according to the urban population was constituted of 11.4% and has risen to 28.53% in the year 2001. The recent census of the year 2011 reflects the vigorous escalation in urban population as it had crossed the 30% mark and reached to 31.16% (Balk *et al.*, 2019). According to the United Nations, (2018a), the number reached 34% in 2018, and it is estimated that it will rise to 40.76% by the year 2030.

The quest of urbanization of the Indian demography is fueled by the industrialization and economic growth (Wang *et al.*, 2018). The service sector has also played a vital role in the economic development of the country. Majority of the amenities including factories, institutes, colleges, schools, hospitals etc. were established in the urban areas and caused large scale migration of people from rural areas to urban dwelling in search of employment, education, healthcare facilities and to live a good quality life. This sudden intrusion of the immigrant population in the urban areas has eventually led to the scarcity of resources and amplified the inflation that triggers a higher cost of living. Also, the Municipal Corporations and Authorities have failed to play their role. As in the majority of the cities/towns, for example, Delhi and Mumbai, the state of the field of municipal waste management are quite miserable mainly due to the lackscadial approach practised by the ULBs (Bhat *et al.*, 2018)followed by recyclable (17-18%. Thus, mostly they were failed to match the emerging needs of the proliferating cities, and the competent reforms in the policies and implementation in the waste management system are the need of the hour.

The recent data released by MoHUA, (2018)·about 1,43,558 MT/D of MSW is generated in the 35 states/UTs and out of which only 24.80% of MSW is scientifically processed. And most of the cities practice open dumping of the comingled waste in the open dumpsites located in the outskirts of the cities and has become breeding grounds for scavengers and mosquitoes and possesses constant threats of hazards like contamination of groundwater and neighbouring water bodies through leakage of leachate (Mandpe *et al*., 2019). These dumpsites are like active time bombs as they might abode radioactive and explosive substances which might explode soon. There are frequent incidents of fire breakouts in the dumpsites resulting in the accelerated release of Green House Gases (GHGs) and air pollutants like methane, SOx, NOx, and particulate matter into the atmosphere thereby deteriorating the air quality of the city (Zhang *et al*., 2019). Figure 1 and Figure 2 represents the infamous events of fire breakout in Bhalswa and Deonar landfills.

Fig. 1: Fire breaks out at Bhalswa Landfill, Delhi (The Hindu, 2017)

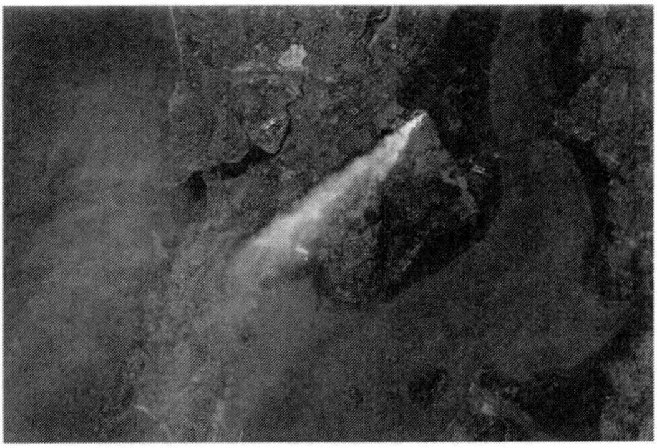

Fig. 2: Satellite image of a weeklong fire break at Deonar Landfill, Mumbai (NASA, 2016)

In recent times the GoI has shown its commitment towards the ignored state of the waste management and has come up with waste-specific rules that promote Public-Private Partnerships powered by the ease of doing business through single-window clearances of the required approval and authorizations from the various government departments and authorities.

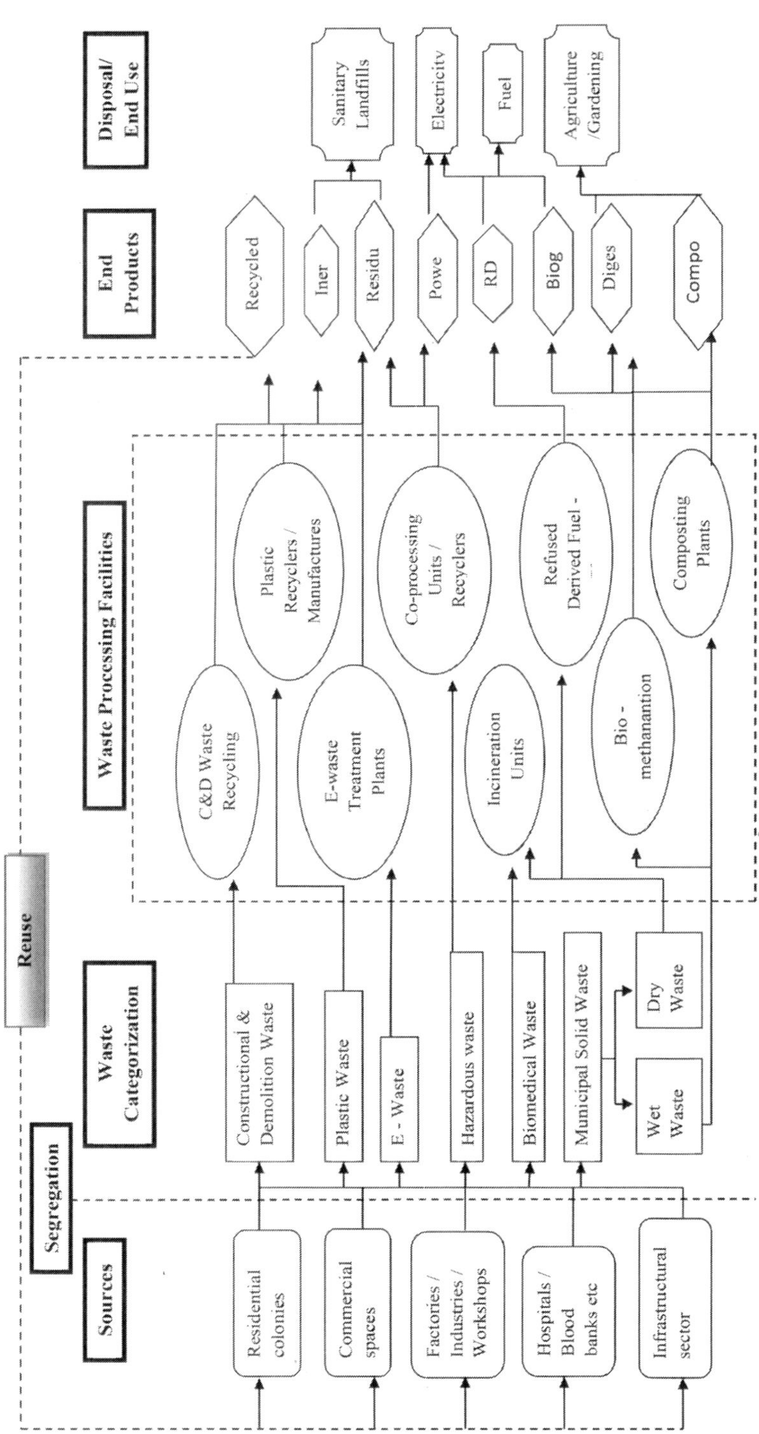

Fig. 3: Recommended integrated Solid Waste Management Plan

To tackle the ramshackle waste management system, the GoI has divided the waste into six categories, namely, plastic waste, e-waste, biomedical waste, solid municipal waste, hazardous waste and construction and demolition waste. And has enacted the waste specified laws to channelize the waste management system in an environment-friendly and sustainable way.

A pictorial representation of an ideal waste management system plan that fulfils enacted legislations in the country and works on the line of minimum disposal of waste in landfills, thereby fulfils the criteria of environmental sustainability.

However, the implementation of these laws is quite unsatisfactory and require a lot of improvement especially in the incorporating the behavioural change among all the stakeholders covering the waste generators, waste pickers, private firms, government bodies, non-government organizations (NGOs), engineers, scientists etc. and on-ground execution of the provisions and rules formulated in the respective acts. The critical targets formulated by the GoI in various strategies, action plans are laid out in the following points:-

1. To make mandatory for every municipalities and ULBs to maintain a comprehensive record for waste generation and composition in their notified area as the quantity and composition of MSW plays a vital role in determining the appropriate waste treatment technology for the given fraction of MSW. As on the year 2019, only a few surveys have been carried out on waste quantification and characterization in the country, and there is lack of reliable and scientific data on waste available for planning the appropriate waste management system.
2. Absolute Implementation of the following waste management practices by all municipalities/ULBs in their respective notified areas:-
    a. Segregation of waste at source by all waste generators and prevention of mixing of segregated waste throughout the waste management cycle, i.e. at the collection, transportation, and processing by following ideal waste management practices by all concerned staffs and personals.
    b. Door-to-door collection of waste by all municipalities/ULBs throughout their notified area through NGOs/Contractors/SHGs.
3. Transportation of collected MSW in covered or closed and well-maintained transits movers like pickup trucks, dumper placers etc. all the vehicles should utilize for at least in two shifts to efficient usage of the transportation resources.
4. Strategically placing of the community bins in all urban localities employing remote sensing technologies like GIS and GPS. So, that

all the community bins are safeguarded from Not In My Backyard (NIMBY) syndrome and should be designed in such a way that they do not attracts stray cattle, street dogs and rodents. Along with the regular collection of waste to avert the foul smell of the degrading MSW.

5. Formalization of informal recyclable waste (scrap) market to optimize the calorific value of the waste stream of the combustible fraction of MSW (>1500 KJ/kg) for round the clock functioning of the Waste to Energy (WtE) plants and thereby creating a self-sustaining environment for WtE plants.
6. Rehabilitation and conversion of all the open dumpsites into scientific landfills and adoption of mitigation measures for groundwater contamination, reduction of GHGs emissions and elimination of open fires epically during the summer season.
7. Preparation of a dedicated long term action plan for waste management system by all municipalities/ULBs for their respective notified areas.

An indicative action plan covering all the waste management activities for ULBs and municipalities based on the quantum of MSW generation elaborated in the table given below:-

Table 1: Indicative Action Plan for Solid Waste Management for various waste generation rate

S.N.	MSW Generation > (tonnes/day)	>500	100-500	50-100	<50
1.	Requisite Approvals	Must have all the obligatory authorizations from the SPCBs/PCCs.	Must have all the obligatory authorizations from the SPCBs/ PCCs.	SPCBs/PCCs may formulate the requirement-centric authority framework.	Approvals may be applied in a phased manner.
2.	Waste Collection /Segregation / Transportation	Ensuring door to door collection of segregated waste and its transportation in the covered vehicles in compliance with the Solid Waste Management Rules, 2016 within two years.	Ensuring door to door collection of segregated waste and its transportation in the covered vehicles in compliance with the Solid Waste Management Rules, 2016 within two years.	Ensuring door to door collection of segregated waste and its transportation in the covered vehicles in compliance with the Solid Waste Management Rules, 2016 within two years.	Ensuring door to door collection of segregated waste and its transportation in the covered vehicles in compliance with the Solid Waste Management Rules, 2016 within two years.
3.	Processing of waste	• Application of unified waste processing technologies, as not a single technology is sufficient to handle the dissimilar waste. • Waste processing facilities should be set up in compliance with Schedule II of Solid Waste Management Rules, 2016.	• Application of unified waste processing technologies, as not a single technology is sufficient to handle the dissimilar waste. • Waste processing facilities should be set up in compliance with Schedule II of Solid Waste Management Rules, 2016.	• Considering the technical and economic feasibility of the ULB, composting, and biomethanation could be utilized.	• Composting could be utilized as the most favourable option and may be supplemented with biomethanation to realize energy demand.

S.N.	MSW Generation > (tonnes/day)	>500	100-500	50-100	<50
4.	Disposal of waste	• Residual waste generated from the waste processing facilities and biomedical waste residue from the incarnation plants should be disposed into the sanitary landfills. • Sanitary landfills must be designed; setup operated as per the Schedule 1 of Solid Waste Management Rules, 2016.	• Residual waste generated from the waste processing facilities and biomedical waste residue from the incarnation plants should be disposed into the sanitary landfills. • Sanitary landfills must be designed; setup operated as per the Schedule 1 of Solid Waste Management Rules, 2016.	• Simpler and easy to execute landfills option might be explored.	• More straightforward and easy to execute landfills option might be explored.

4.3 Legislation and Laws on Waste Management in India

The decade of the 1970s can be termed as a decade of awakening for the Indian economy as during that period India was experiencing major up-gradation in agriculture powered by the green revolution and the slogan of 'Garibi Hatao' means remove poverty was quite prevalent in the politics and among the masses (Planning, 2017). In the year 1972, the breakthrough had taken place in the field of preserving and protecting the environment as the Parliament of India (PoI) enacted the 'Wildlife Protection Act, 1972' that embraces wildlife (flora and fauna) in its natural habitat and niche. This act allows state governments to transform the ecologically sensitive areas into wildlife sanctuaries and national parks (Forester and 2017, 2017). In the subsequent years, PoI has come up with several laws, legislation and rules to strengthen the movement of persevering the environment and nature by keeping the concept of waste management in the mainframe. As India was among 122 signatories nations at the United Nations Conference on the Human Environment that was held in Stockholm, Swed(Jain, 1984)en in June, 1972 and has pledged to embrace the 26 principles concerning the environment and development in the national policies and governance (Handl *et al.*, 2012).

In the year 1974, the GoI had enacted the Water (Prevention and Control of Pollution) Act, 1974 that leads to the foundation of various regulatory institutions and testing laboratories like Central Pollution Control Board (CPCB), State Pollution Control Board (SPSB), Central and State Water Laboratories for compliance of execution of the provision elucidated in the rules and has established standards for water quality and effluent (Jain, 1984). Under this act, polluting industries must ask for permission to discharge waste effluents into the water bodies. On March 29th 1981, the GoI broadened the horizons in the pursuit of sustainable development by enacting the Air (Prevention and Control of Pollution) Act, 1981 that empowers the Central and State Pollution Control Boards to implement the provision of the rules and penalize the defaulters. It also leads to the establishment of Central/State Air Laboratories to monitor and analyze the samples for pollutants and particulate matter

The year of 1986 witnessed the rise of the concern of government towards the environment as on May 23rd 1986 the GoI has enacted the Environment (Protection) Act, 1986, which eventually becomes an umbrella act for the subsequent laws in the field of waste management (Department of environment, 1986). Horrifying aftermath of Bhopal gas tragedy that has led to 15,000 deaths has exposed incompetence in the legislation of the country against the industrial hazards (Bisarya and Puri, 2005). It necessitates GoI to formulate strict laws and rules that will prevent accidents due to mishandling of hazardous substances and provide legitimate compensation to victims and their dependents. This act empowers the centre and state governments to take all the necessary measures for the protection of the environment and has laid the foundation of a nationwide programme for the prevention, control and reduction of environmental pollution. Various environmental laboratories were set up under this act for analyzing the samples of soil, air, water and related substances. Formulation of standards to check the discharges and emissions released from any industries/individuals/operations/process has been done (Department of environment, 1986). Induction of 'Polluter Pays Principle' was made in the act which states that whosoever responsible for the degradation of the environment had to pay the expenses entailed to restore it (Department of environment, 1986).

Within a short period of three years, the GoI had come up with the Hazardous Wastes (Management and Handling) Rules, 1989 that defines the hazardous waste based on characteristics of the trash that possibly possess threats to the environment or health of the handler of waste (Dutta *et al.*, 2006). Also, responsibilities of producers, occupiers, importers and exporters of hazardous wastes for its collection, packaging, labelling, transportation, treatment, storage and disposal were formalized in the act. The central/state pollution control boards are vested with the power to authorize and to suspend the enduring authorizations in case of violations of rules by the proponents (MOEF, 2016).

In the year 1991, the Public Liability Insurance Act was enacted that provides insurance cover against the accidents caused due to mishandling of hazardous waste for any individual other than workers. This act made compulsory for every owner, board's members, and directors to take out insurance policies against all possible mishaps before setting up the facility and must renew it regularly. Also, a medical cover of up to Rs. 12,500 per victim with an additional relief of Rs. 25,000 to the victim in case of a fatal accident and a monthly relief up to Rs. 6,000 for a maximum period of three months in case of loss of earnings was permitted (Of, 1992). In case of violation of the rules, there is a provision of both imprisonments from 3 months to 7 years and fine from Rs. 1,000 to Rs. 1,00,000 for the defaulters.

On June 17th 1995, National Environment Tribunal was established under the National Environment Tribunal Act, 1995 for speedy disposal of matters related to compensation for the damages and adversities arose from the accidents caused due to mishandling of hazardous waste. The compensation amount should be credited to the Environmental Relief Fund and should be spent as prescribed in the act. In case of breach of the order made by the tribunal, the offender shall penalize with extended imprisonment of three years, or fine of Rs. 10 lakhs, or both (Central Government, 1995).

The Biomedical (Waste Management & Handling) Rules, 1998 was enacted to formalize the biomedical waste generated from the medical and health care units by defining the biomedical waste, identifying the waste generators and setting out the responsibilities of waste generators, occupiers and handlers in carrying out activities like segregation, packaging, transportation and storage. Prescribing authorities were set up in every state and union territory to monitor the execution of the rule and to provide necessary authorization to the stakeholders. This act was succeeded by the Bio-medical Waste Management Rules, 2016 (Rules *et al.*, 2019).

Eventually, GoI had come up with the Plastics Manufacture, Sale and Usage Rules, 1999 to regularise the manufacturing and consumption of plastics especially in packaging and carrying the various commodities in the form of carrying bags. The minimum thickness of the carry bags was made 20 microns and was raised to 40 microns in the superseded version of the act in the year 2001 and then to 50 microns in the Plastic Waste Management Rules, 2016 (Singh and Sharma, 2016).

The most noteworthy step taken by the GoI in the turf of waste management is the enactment of Municipal Solid Wastes (Management and Handling) Rules, 2000 that are aimed to formulate and implement the waste management rules in the area under the jurisdiction of municipal authorities (Patel *et al.*, 2010) generated from different activities in the township and city areas is a subject of deep concern for its proper management. The improper management of the MSW is a major cause for water, air and soil pollution. The population explosion and sustained drive for economic progress and development have resulted in a remarkable increase/ change in quantity and characteristics of MSW generation over the last 20 years. The local bodies are responsible for the management of the MSW in the State. The Municipal Solid Waste (Management & Handling. These rules mandate the incorporation of all the available waste treatment technologies like composting, bio-methanation, incineration, refuse-derived fuels (RDF) etc (Patel *et al.*, 2010)generated from different activities in the township and city areas is a subject of deep concern for its proper management. The improper management of the MSW is a major cause for water, air and soil pollution. The population explosion and sustained drive for economic progress and development have resulted in a remarkable increase/ change in quantity and characteristics of MSW generation over the last 20 years. The local bodies are responsible for the management of the MSW in the State. The Municipal Solid Waste (Management & Handling.

In the year 2016, the massive transformation of existing acts and inclusion of new laws has been observed. These acts were incorporated to fill the existing loopholes of the waste management system practised in the country and deals directly in the enforcement of waste management. As per the GoI waste can categorize in six categories, namely plastic waste, e-waste, biomedical waste, hazardous waste, C&D waste and municipal solid waste. A brief outline of the legitimate rules in the field of waste management is described in the table below:-

<table>
<tr><td>Batteries (Management and Handling) Rules, 2001</td><td>Date of Notification: May 16th 2001</td></tr>
<tr><td colspan="2"><ul><li>Applicable to all the stakeholders dealing in the sale, purchase and use of batteries or its components.</li><li>Enforcement of prescribed guidelines on the sale and purchase of new and old batteries.</li><li>Ensuring the safe collection, transportation and recycling of used batteries.</li><li>All dealers have to give an appropriate discount on every used battery to the customer and ensure that the collected batteries are of similar specifications that of sold ones.</li><li>Collected used batteries should be sent only to authorized recyclers.</li><li>All the importers of the batteries have to register with the Ministry of Environment and Forest.</li><li>CPCBs/SPCBs were made prescribed authority to check the compliance of the rule at centre and state levels respectively.</li></ul></td></tr>
<tr><td colspan="2"><ul><li>Organizing public awareness campaigns on the hazard of lead, recycling of batteries and location of designated collection centres through advertisements, publications and posters.</li></ul></td></tr>
<tr><td>Hazardous And Other Wastes (Management and Transboundary Movement) Rules, 2015</td><td>Date of Notification: July 24th 2015</td></tr>
<tr><td colspan="2"><ul><li>Superseded the Hazardous Wastes (Management, Handling and Transboundary Movement) Rules, 2008.</li><li>Applicability of the rules extended by the inclusion of ‘other waste’ with hazardous waste.</li><li>Simplification of the procedure by merging all the approvals through single window clearance. Aimed to bring swiftness in the implementation of the rules.</li><li>Co-processing has been identified as a superior method over disposal as it can be exploited for energy recovery.</li><li>Elaborates the responsibilities of occupiers and state governments for eco-friendly management of hazardous and similar waste.</li><li>Several hazardous items like metal scrap, paper waste, and various electrical and electronic items were exempted from the need for authorization and permission.</li><li>State Pollution Control Boards are vested with the power to suspend or cancel an authorization in case any violation of the rules by an occupier.</li><li>Occupiers have to maintain a record of all the activities concerning with the handling of hazardous wastes like sale, storage, recycling, utilization etc. and may store hazardous and other waste for up to 90 days.</li></ul></td></tr>
</table>

Plastic Waste Management Rules, 2016	Date of Notification: March 18th 2016
• Succeeded Plastic Waste (Management and Handling) Rules, 2011. • Minimum thickness of plastic bags is raised from 40 to 50 microns. • Expansion of the jurisdiction of the rules with the inclusion of rural areas along with urban areas. • Utilization of plastic waste in the construction of flexible pavements for highways and districts roads. • For the first time, the manufacturers have given the responsibility to plan and design the plastic waste management system in collaboration with the LAs. • Fine was imposed on the vendors and shopkeepers who are using plastic carry bags that do not conform to the rules. • CPCBs have been directed to formulate guidelines for thermo-set plastics (plastics difficult to degrade). • Prohibition of charging fees for carrying bags.	
E-Waste (Management) Rules, 2016	Date of Notification: March 23rd 2016
• Superseded the E-Waste (Management and Handling) Rules, 2011. • Expansion of the applicability of the rules to dealers, refurbishes and Producer Responsibility Organizations (PROs) and inclusion of Compact Fluorescent Lamp (CFL) and other mercury-containing lamps under the purview of rules.	
• Collection centers must be set up by the producers; otherwise, they should arrange a suitable buyback mechanism. • CPCBs are entitled to provide single authorization for producers for swift implementation. • Responsibilities of concerned government bodies like the department of industry, labour, ULBs were devised for sound pan India implementation.	
Bio-medical Waste Management Rules, 2016	Date of Notification: March 28th 2016
• Superseded the Bio-Medical Waste (Management and Handling) Rules, 1998. • Applicability of the rules has been expanded with the inclusion of vaccination camps, blood donation camps, surgical camps and any other healthcare activity. • Prohibition on the use of chlorinated gloves, blood bags and plastic bags. • Onsite treatment or sterilization of BW generators as per standards set up by the WHO. • Incorporation of Bar-Code System on the bags or containers used in BW disposal. • Classification of BW is now limited to four categories instead of 10 categories for competent segregation of BW. • State Governments have to allocate a suitable place of land for setting up of standard BW treatment and disposal facilities.	

<table>
<tr><td>Construction & Demolition Waste Management Rules, 2016</td><td>Date of Notification: March 29th 2016</td></tr>
<tr><td colspan="2">
• Applies to everyone who generates construction and demolition waste.
• C&DW should be segregated into four streams, i.e. concrete, soil, steel and wood, plastics, bricks and mortar.
• State governments should develop waste processing and recycling sites.
• Mandatory utilization of 10-20% C&D wastes in the government constructional projects.
• Assigning the duties of central and state authorities to provide the requisite environment for swift implementation of the rules like:-
1. Bureau of Indian Standards (BIS) entitled to prepare a code of practice and standards for products made from C&D waste.
2. All the Local Authorities (LAs) has to come up with a robust framework of policies and infrastructure the supports the implementation of the rules.
3. LAs should provide appropriate incentives to waste generators for on-site salvaging, recycling and utilizing of C&D waste.
</td></tr>
<tr><td>Solid Waste Management Rules, 2016</td><td>Date of Notification: April 8th 2016</td></tr>
<tr><td colspan="2">
• Replaced the Municipal Solid Wastes (Management and Handling) Rules, 2000.
• The ambit of the rules has been expanded to urban agglomeration, census towns, defence establishments, religious place and areas under Indian railways, airports, airbase, ports and harbour.
</td></tr>
<tr><td colspan="2">
• Compulsory segregation of waste at its source by the generators to streamline waste to wealth by recovery, reuse and recycle.
• Integration of waste pickers/ rag pickers and waste dealers/ kabadiwalas into the formal system through state governments.
• Littering and burning of waste in open areas, streets, roads and drains are prohibited.
• All waste generators have to pay 'User Fee' to waste collector.
• Provision of 'Spot Fine' for not practising segregation of waste and littering in open areas.
• All Bulk Waste Generators (BWGs) have to practice segregation of waste and should set up waste treatment facilities (composting and bio-methanation) in their premises for wet waste and should hand over the valuable dry waste to authorized waste pickers and recyclers.
• Allocating at least 5% of total area minimum of five plots for setting up of recovery/ recycling units in Special Economic Zone (SEZ) and industrial areas.
• All manufacturers of disposal products have to provide finical assistance to LAs for developing the waste management system.
• All the industrial units, including cement and thermal power plants located within 100 km radius of the RDF plant, have to use at least 5% of the total fuel requirement.
</td></tr>
</table>

- Events and gathering of more than 100 persons should practice segregation of waste.
- Only the non-combustible, inert and residual should be disposed of in sanitary landfills.
- In hilly areas instead of constructing landfills, transfer stations should be utilized to collect residual waste from the waste treatment plants and inert waste. The collected waste should be transfer to the sanitary landfill located within 25 km in the plain or downhill area.

4.3. Municipal Solid Waste Management Rules 2016

Municipal Solid Waste Management (MSWM) broadly deals in the collection, handling, treatment and disposal of MSW such that there will be least environmental and social impacts induce by it and thereby creates a clean and hygienic environment for all inhabitants. The Municipal Solid Waste Management Rules, 2016 was enacted by GoI on April 8th 2016 and drove powers from sections 3, 6 and 25 of the Environment (Protection) Act, 1986. These rules have superseded the Municipal Solid Waste (Management and Handling) Rules, 2000. The Municipal Solid Waste (Management and Handling) Rules, 2000 have failed drastically as even after the 15 years since its inception only the 90% fraction of MSW was collected, out of which about 27% of waste was treated and a negligible amount of 3.2% waste has found its way to scientific disposal site (CPCB, 2015).

The primary cause of the failure of these rules was that ULBs and municipalities are failed to match up the services, facilities and infrastructure required for efficient waste management of the behemoth amount of MSW generated by the urban localities (Guerrero *et al.*, 2013).

This pathetic and wretched state of the MSW system has compelled GoI to come with a robust set of rules that can counter and resort the limitations of the existed rules and regulations. Some of the salient features of the SWM Rules, 2016 are listed below:-

- Integration of SWM rules with SBM is done by making it compulsory for bulk waste generators (BWG) to practice segregation of waste into wet and dry waste along with the treatment of wet waste through composting within their premises.
- The applicability of SWM Rules is raised from municipal areas to urban agglomerations, industrial townships, religious and historical places.
- Segregation of waste and transportation of segregated waste to the treatment facility is compulsory.

- Integration of informal sanitary waste workers into a formal system through Municipalities and Self Help Group.
- Littering and burning in public places is prohibited.
- Collection of 'user fee' and 'spot fine' for waste collection and littering is introduced.
- 5% of the total area should be allocated for waste processing/recycling in Special Economic Zone and industrial estates.
- All manufacturers of disposable and packing products have to provide financial support to municipalities for effective waste management of their products.
- All thermal plants and industrial units located within a 100km radius from RDF plant should incorporate at least 5% of their fuel requirement through RDF.
- Construction and demolition waste (C&DW) should be managed as per C&MW Rules, 2016.
- All public gatherings with more than 100 persons have to ensure segregation of waste at source.
- Special provisions for SWM in hilly regions were made like avoiding the construction of the landfill, establishing a waste transfer station within 25 km at plain terrain.
- Responsibilities of all the major stakeholders, including waste generators, all the concerned ministries like MoEF&CC, MoHUA, MoC&F, MoA, MoP, MNRE, CPCBs, SPCBs, etc. were fixed.
- Principles and standards for Waste to Energy (WtE) process, sanitary landfills, processing and treatment of solid waste like composting, incineration, etc. were formulated.

4.4. Initiatives are taken by the Government of India

4.4.1. Swachh Bharat Mission

On September 24th, 2014 India has become one of the four elite nations in the world after the USA, Russia, (European Space Agency) ESA that has successfully placed a spacecraft in the Martian orbit (Sundararajan, 2013)2012, the 66th Independence Day of India, Prime Minister Dr. Manmohan Singh formally announced Mangalyaan, the nation's first interplanetary mission to Mars. The Mars Orbiter mission was initiated by the Indian Space Research

Organization (ISRO. However, on the ground, India was lacking in providing the basic sanitation amenities to its citizens, as in the year 2014 only 38.70.% of the people had access of toilets and the rest of the population practiced open defecation (Sambodhi Research and Communications *et al.*, 2015). As per Fewtrell and Colford, (2005) poor sanitation is directly involved in the spreading of diseases like diarrhea, dysentery, hepatitis A, typhoid and polio, especially in the tropical countries. Consequences of the poor sanitation and hygiene are quite horrifying as it has led to 9.2% of the total deaths and 31.4% of the total deaths in below 5 years age group in the country along with the loss of INR 1,316 billion in year 2006 to the Indian economy (DeFrancis, 2011)180 (US$48.

Moreover, the issue of improper or lack of sanitation has been raised at global platforms too, and concrete steps to resolve this issue has been taken from time to time. In the year 2000, Millennium summit was held at UN Headquarters, New York and 8 Millennium Development Goals (MDGs) were finalized with a target '7c' under the goal '7' to achieve 50% reduction in the population that deprived of having sustainable access to safe drinking water and basic sanitation by the year 2015 (World Health Organization, 2005). Apart from this in September 2015, Agenda 2030 for Sustainable Development Goals (SDGs) was adopted by world leaders at the United Nations, which calls on countries to begin with sustainable efforts for achieving 17 SDGs over the next 15 years, including a goal to "ensure the availability and sustainable management of water and sanitation for all (Herrera, 2019).

India has shown a strong will for achieving these goals and to become a clean and hygienic country. On October 2nd 2014, GOI launched SBM in two variants, i.e., SBM urban and SBM rural throughout the length and breadth of the country with the following cumulative objectives (Vishwakarma, 2016):-

- Eliminating open defecation practice and improving the general quality of life.
- Speed up the sanitation coverage throughout the country.
- Inculcating scientific and sustainable sanitation practices with its linkage with public health at the grassroots level through awareness campaigns.
- To encourage cost-effective and suitable technologies for ecologically safe and sustainable sanitation.
- To create a favourable ecosystem for the private sector participation through financial support by the local bodies.

GOI has shown a commendable interest to make sanitation as a national goal and has made SBM a mass movement by involving the state governments, municipalities, private parties, Non-governmental organizations (NGOs), resident welfare association (RWAs), panchayats in unison for a noble cause of achieving clean and ODF free India by October 2nd 2019 to bestow a tribute to Mahatma Gandhi on his 150th birth anniversary (Vishwakarma, 2016). About 28 states and union territories have become Open Defecation Free ODF, sanitation coverage has gone up from 38.70% to 99.17%, and 9.27 crores of toilets were built throughout the country as on May 27th 2019 with the significant part of funding coming from the central government (Vishwakarma, 2016). Moreover, it has become a "people's movement". It has impacted deep inside the rural interiors of the country by addressing the orthodox attitude of rural households for open defecation and its health and economic impacts. SBM has got remarkable success in behavioral change for open defecation (Vishwakarma, 2016).

According to Sampath, (2019) more than 3,00,000 lives would have saved by the time SBM ended in 2019. However, there are still some lacunas in the programme like there is no mention of elimination of casteism in the objectives of SBM as it is still prevalent in rural society and only the lower sections of society are indulged in the cleaning and scavenging works (DeFrancis, 2011)180 (US$48. They often also don't get any recognition of their service. Also, the construction of the toilets is done hastily, as to achieve the gargantuan target in a short timeframe. The basic requirements for the toilets were side-lined like site selection, groundwater table level, the strength of concrete, quality of building materials. Even though SBM has few lacunas, still SBM is India's most prominent campaign which is successful in achieving its set objectives within a stipulated timeframe. Its most significant achievement is inducing behavioural changes regarding cleanliness and sanitation practices in society (Mohapatra, 2019).

4.4.2. Role of the Urban Local Bodies in Urban and Semi-urban parts

ULB is the principal authority that is responsible for the successful implementation of MSW Rules, 2016 throughout its notified area. The main areas on which a ULB should be focusing on to make its territory clean and free from the menace of filthy litter and garbage are the development of integrated MSWM system, robust and leak-proof sewage system, maintaining cleanliness at all public and commercial places including roads, streets and railway tracks. Even in the SBM-Urban, the key emphasis is given to strengthening the MSWM system of the city. MSWM is broadly comprised of four stages, namely, collection, transportation, treatment, and disposal. As per the Indian scenario, most of the ULBs have a lot of room to improve under this aspect of SBM. By incorporating the following practices, a ULB can take a step forward in achieving the clean and hygienic habitat for its citizens: -

- 100% door to door collection (DTDC) of segregated waste should be practised, and the defaulters should be fined accordingly.
- The collection of waste should be done once in a day at residential places and twice in a day at all commercial places to prevent the overflowing of bins and decomposition of waste that can produce odours.
- All the BWGs should be identified, and treatment of OFMSW generated by them should be done on their premises only.
- The collected MSW should be transported in well-maintained vehicles and shall be covered to prevent odours and spillage of waste.
- If the distance from the collection point to the treatment/processing facility is more than 25 km, in that case, a waste transfer station should be established.
- GPS monitoring of all waste collection vehicles should be done along with route optimization for efficient usage of the vehicle and fuel savings.
- The dry waste should be recycled, and non-recyclable dry waste can be incinerated in WTE plants.
- The high calorific value dry waste can be transformed into RDF and can be co-fired with coal in thermal power plants, cement manufacturing units, brick kilns etc.
- The C&D Wastes should be recycled, and different products such as bricks, blocks can be made through C&D Wastes.
- The wet waste should be treated through decentralized composting units and turned into compost which can be applied as a fertilizer in the agriculture fields and also for recreational purposes.
- Wet waste can also be anaerobically decomposed to produce biogas, and the digestate can be used as a fertilizer similar to compost.
- Sanitary landfills should be constructed for disposal of only inert, non-biodegradable, non-combustible, non-hazardous waste and waste generated from the waste treatment/processing facilities.
- All open landfills should be converted into sanitary landfills, and if not possible, it should be appropriately capped according to MSW Rules, 2016.

- Integration of all informal waste pickers into organized, sanitary workers through ULB, SHGs, and NGOs.
- All sanitary workers should wear PPE and should be vaccinated regularly.

Sewage network system should cover all parts of the city, and all public and household toilets should be connected to the sewage system. The sewage treatment should be capable of treating all the wastewater of the city, and the disposal of sewage sludge should be done scientifically (Paliya *et al.*, 2019).

Apart from this, cleanliness should be maintained in all public and commercial places of the city by installing garbage bins having different compartments for wet and dry waste at every 500m distance. Street sweeping should be practised in all commercial and residential places twice in a day and once in a day, respectively. All major roads should be cleaned by using roads cleaning equipment during off-peak hours.

Beautification of old parts city and flyovers should be done through graffiti, paintings, artworks etc. All the historical and culturally important places should be maintained clean. Moreover, littering at public places should be made strictly prohibited, and defaulters should be fined heavily.

4.4.3. Role of Gram Panchayats in Rural parts

The major habituated landmass in India still comes under the rural area that accumulates 70% of the total population of the country and accounts 83.3 crores of persons according to the latest census of 2011 (Chandramouli, 2011). The GoI defines the rural area that fulfils the criteria of the population size of less than 5,000 with the population density of not more than 400 per sq km, and at least 25% of the male population took agricultural as their main profession (UNICEF, 2011). The state of MSWM system and sanitation in the rural areas of the country was quite miserable as compared to the urban areas (Ghosh and Cairncross, 2014).

Although, the several sanitation programmes and cleanliness drives were being organized throughout the country since the inception of independent India. The first rural sanitation program was implemented in the year 1954 under the first five-year plan of GOI. But the census of the year 1981 reveals that only 1% of sanitation coverage was achieved in rural India till the year 1981 and validates the collapse of the rural sanitation programme, 1954 (Sastry, 1999). Further, GoI has launched several sanitation programmes sporadically like Central Rural Sanitation Programme (CRSP) in 1986, Total Sanitation

Campaign, 1999, Nirmal Bharat Abhiyaan in the year 2012. The recent and most remarkable up-gradated sanitation campaign is SBM that was launched in the year 2014 to achieve universal sanitation coverage within a stipulated time of 5 years (Mehrotra and Ghosh, 2013).

In India, Gram Panchayats are the grassroots level governance bodies of rural India which came into existence in the year 1992 through the 73rd amendment of the Indian constitution (Nambiar, 2001). To make SBM–Gramin, a successful story Gram Panchayats has to play a responsible and robust role especially in changing the behaviour of rural Indians towards sanitation along with constructing and using toilets in their household premises.

Gram Panchayats should take the following practices in its functioning for amplifying the endeavors for establishing a robust MSWM system suited for rural requirements: -

- (Information, Education and Communication) IEC should be used as a tool to trigger the sanitation requirement in the rural habitats by creating awareness for the sanitation, hygiene in the day to day life and its advantages like prevention of diseases, improvement in productivity of occupants and establishment of a healthy environment to sustain a livelihood.
- Interpersonal communication through a door to door contact can play a vital role in establishing effective vertical communication between the decision-makers and the end-users. Thereby, it creates a room for the policymakers to induce the required corrections in the formulation and implementation of policies. At the same time, interpersonal communication with end-users also enhances the favourable outcomes of the policy.
- The construction of toilets should be done following building standards and must connect to either a sewerage network system or dual-pit septic tanks.
- All the toilets should have water availability.
- All of the solid and liquid waste generated from the village should be collected, transferred, treated and disposed of through the cheap, robust and locally available waste treatment practices.
- The segregation of the solid waste into wet and dry fractions should be done efficiently.

- Treatment of wet solid waste can be done through composting, vermin-composting and bio-methanation in decentralized treatment facilities.
- The wet solid waste should be recycled locally or can be transported to the nearest RDF processing plant and latter to the sanitary landfills.
- The liquid waste should be collected in closed sewerage network. It can be treated through waste stabilization pond technology, duckweed based wastewater treatment, phyto-rid technology (developed by CSIR-NEERI) and anaerobic decentralized wastewater treatment technique.
- All Schools, Aganwadi Kendras, Sanitary Complexes, Public places should have separate toilets for the two genders within their premises.
- All women toilets should be equipped with the sanitary pad vending machines accompanied with proper disposal mechanism for the used sanitary pads.

4.4.4. Role of other parts of the society in Municipal Solid Waste Management

The term society originated from the Latin word *'societas'* that means a friendly and civilized interaction between the groups of individuals. The great Scottish economist, Adam Smith, defines society as, the existence of multiple groups of individuals without any mutual love or affection and only they refrain from causing harm to each other. The society acts as a Keystone for the successful implementation of a municipal waste management system (Kumar *et al.*, 2009)which are facing a serious pollution problem due to the generation of huge quantities of solid waste. This paper presents an assessment of the existing situation of municipal solid waste management (MSWM. Whether it is a waste generation or its disposal, it is the society who has to face the consequences or to gain the benefits (Khandelwal *et al.*, 2019).

All kind of waste generators like households, vendors, merchants, users at commercial and public establishments like cinema complexes, stadiums, railways stations, airports, schools and medical care units must have to play their role in the MSWM.

All the sections of the society must have to develop a sense of responsibility and must fix their accountability over the thrash they generate every day. As per the SMW Rules, 2016, all the waste generators have to practice segregation of waste by keeping different colour bins for wet and dry waste. It would be even better if they practice in-house composting of the wet waste and apply this compost in gardening activities. E-waste and hazardous waste should also be separated and to be deposited in the nearest e-waste collection centre.

GoI has formulated various policies and has enacted all the required legislations along with the establishment of the concerned authorizing bodies like CPCBs, SPCBs, PCC, laboratories etc. it is the duty of ULBs and citizens to implement. It follows these rules and makes their respective localities clean and free from litter by developing a robust municipal waste management system.

By adopting the following practices, various sections of society may complement the making of an efficient and workable solid municipal waste management system:-

- There is a strong need for mindset shift among the waste generators; they need to take ownership of their role in the garbage collection and must practice segregation of waste at its source.
- Waste generators may adopt the concept of 3R's, i.e. Reduce, Reuse and Recycle for solid waste management right from their houses, shops, restaurants etc. this will significantly reduce the quantum of waste generation and also curtails the greenhouse gas emissions, conserves resources and saves energy.

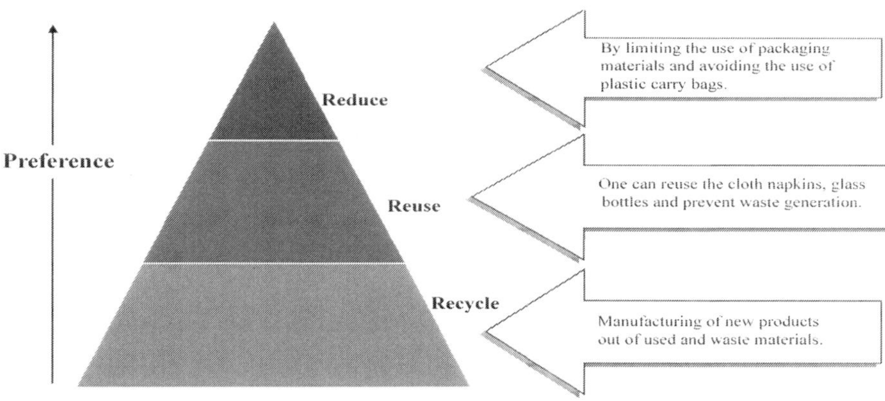

Fig. 4: Concept of 3R's in the MSWM

- As listed in the MSW Rules, 2016, all bulk waste generators, i.e. the establishments that are generating MSW more than 100 kg/day have to segregate their waste into a dry and wet fraction and have to treat the wet fraction of waste in their premises itself. This could reduce the total MSW more than 50% as about 44% to 62% fraction of the total MSW is organic.
- All the citizens should not litter the garbage and thrash on streets and in open drains. As during the monsoon season, the streets and roads get waterlogged due to chocking of the drainage system from the litter and trash.

- NGO's may act as a fourth pillar in solid waste management apart from the government, ULBs, and the citizens. NGO's can act as a vital tool in creating awareness among the masses regarding the ideal practices that need to be followed for effective solid waste management like segregation of waste, in-house composting, NGO's
- Informal waste recyclers could also play a pivotal role in the strengthening of the existing MSWM system especially in low and middle-income group countries, which lacks adequate infrastructure and workforce to deal with an enormous amount of waste generated.
- Public-Private Partnership (PPP) could be utilized as a favourable strategy to prepare models of MSWM concerning individual regions. Especially in the field of collection and transportation of MSW private firms and companies could play a significant role in consistence and economical way.

4.5. Waste Scavengers and Ragpickers

Waste scavenger, waste pickers or rag pickers are unorganized workforce involved in the MSWM system and act as a backbone of the MSWM system of several cities of developing and developed countries. In India, it is estimated that about 1.5 million to 4 million waste pickers are involved in the sorting, cleaning and gathering of recyclable waste and further supplies recyclable waste to the informal waste recycling market (Dandapani, 2018). However, even after playing a riveting role in the MSWM, the working conditions of rag pickers are piteous and unhygienic. According to Chokhandre *et al.*, (2017) the occupation of a rag picker is characterized by 3 D's dangerous, drudgery and demanding. Also, it is found that the possibility of acquiring diseases like respiratory illness, eye infections, and gastric disorders was about to 86%, 61% and 68% higher than the residents residing in the neighboring localities of the landfill respectively (Mandpe *et al.*, 2019a; Singh *et al.*, 2021).

Even the children below the age of 14 years have been employed in the dirty job of extracting out the recyclable waste from the comingled waste dumped in the open dumpsites (Joshi and Ahmed, 2016). Thus, it is evident that it's high time for induction of concrete reforms and implementation of the enacted legislation and rules. It has been observed that the MSW rules, 2016 have entitled the waste pickers for certain benefits like recognizing and authorizing the informal waste pickers into formal MSWM system and employed them into a door-to-door collection of MSW. The state policies and authorities are directed to appreciate the commendable service of the waste pickers in the collection and extraction of a precious recyclable fraction of waste from comingled MSW and thereby releasing the burden of behemoth quantum of

MSW that to be subjected to be dumped into the landfills (Joshi and Ahmed, 2016).

Even the former Environment Ministry had praised the work of rag pickers by stating them as the saviour of the country and even recognized their efforts by granting a national award and a cash prize of Rs. 1,50,000 for three rag pickers and private bodies associated with innovative waste management technologies (Bhattacharya, 2017).

Apart from these initiatives, the following recommendations should be incorporated in the MSWM system to synchronize the incredible quality of the rag pickers into a sustainable MSWM system:-

- Regular health checkups and vaccination camps should be organized mainly for the waste pickers community.
- Strict check to ensure that no child below the age of 14 years is gets employed as a waste picker and in similar jobs.
- Establishment of training institutes and academies for educating and empowering the rag pickers regarding the handling of waste of various categories like hazardous, metals, sharp and pointed objects, biomedical waste etc.
- Involving of waste pickers in the industries and start-ups that work in the recycling and creating best out of the waste like sculptures, artefacts etc. thus creates an alternate source of employment and improves the quality of life of the waste pickers.
- Opening of schools in the vicinity of landfills and encouraging the children and desired rag pickers to attend schools and empower themselves by getting literate.
- Organizing regular awareness campaigns among the waste pickers regarding the importance of personal safety and health equipment at their workplace like the use of gloves, safety jackets, safety boots, safety goggles etc.

4.6. Municipal Solid Waste Management Practices in Developed Countries

The world's first country that has gained the title of a 'developed nation' is the United Kingdom, and the major factor behind this achievement was industrialization as it has fueled the economy and makes the nation prosperous Followed by the United Kingdom, Belgium was the second nation that was declared 'developed'. Recently, in the year 2018, Poland has joined the league of developed countries. Earlier, the industrialization and economic parameters like Gross Domestic Product (GDP), per capita income etc. are used to assess a country's development. However, in recent times several criteria for defining a country's performance in fostering its citizens has been introduced like Human Development Index (HDI), Gross Happiness Index (GHI) etc. that rely on the existing conditions on social, health, education and economic front of the citizens. At the same time, various environmental and ecological indices have been developed and applied to measure the efforts made by a country in compliance of environment preservation and sustainability like ecological footprint, Dow Jones Sustainability Index, Environmental Performance Index etc.

Both genre of the indices that is the one which is economically driven and another one which is environmental driven often contradicts with one other as the Figure.5 below represents the world maps of the HDI and EF respectively.

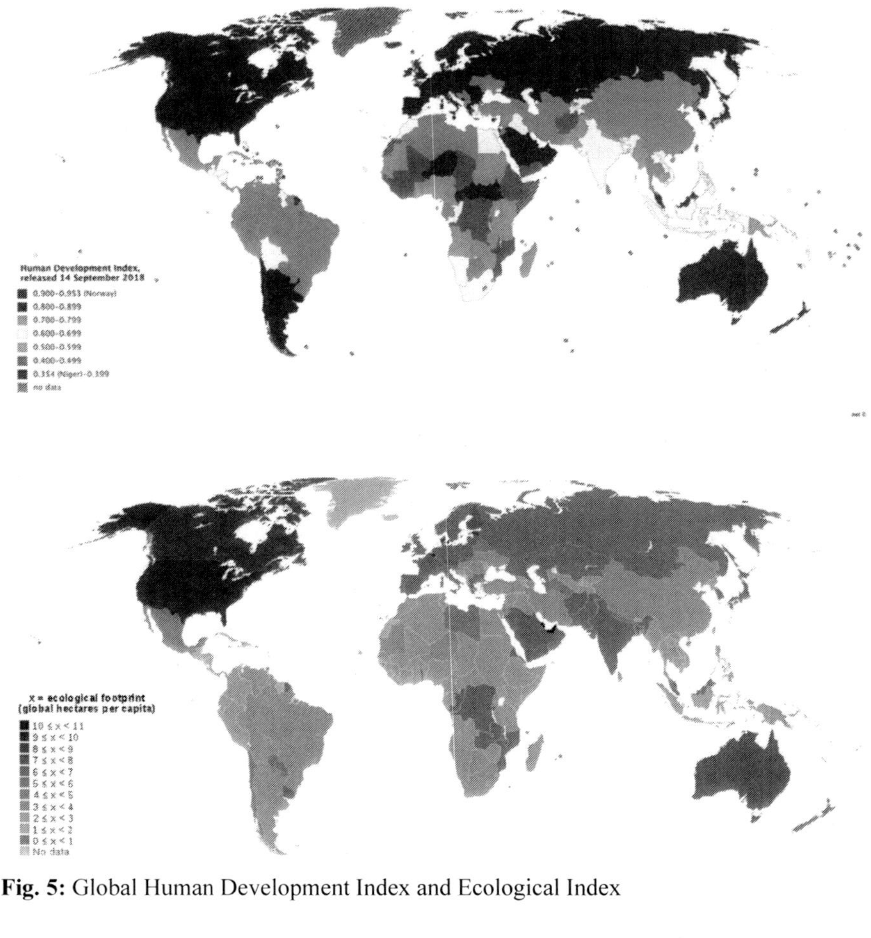

Fig. 5: Global Human Development Index and Ecological Index

By comparing the above maps, it has been observed that the countries which had performed exceptionally well in HDI like USA, Canada, Australia, New Zealand, Belgium had failed miserably in EF parameter. Thereby this exhibits that the so-called development and higher standards of living had laid on the cost of environment and nature.

In compliance with SWM, developed countries have pioneered in the policymaking and implementation of bye-laws and provisions enlisted in the legislations concerning SWM. However, due to the lifestyle and food habits of the people, the per capita waste generation is quite high. As urbanization and development have a positive correlation with MSW generation. Though the quantum of waste generation is quite enormous, these countries have sufficient resources, technology and infrastructure to deal with the behemoth amount of waste generated.

These countries beliefs in the philosophy of zero waste that works on the principle of 'cradle to cradle' approach. The term cradle to cradle is applied in the field of life cycle assessment, and it stands for recycling the product into a new product after the accomplishment of its life cycle.

Comprehensive review of the MSWM system adopted in the capital cities of various developed countries is depicted in Table 2.

4.7. Municipal Solid Waste Management Practices in Developing Countries

According to the International Statistical Institute (ISI), the countries which have Gross National Income (GNI) per capita per year lower than the US$ 12,056 are termed as developing countries. Presently, in the world, there are 133 developing nations. The state of MSWM is quite wretched in these countries as compared with the developed ones. As these nations mostly become independent nations in the mid and latter 20th century. And starts off their journey to become a self-reliant, developed and prosperous nation. However, even in the 21st century, there is still a long path to cover for the majority of the developing nations to meet the basic requirements of their citizens.

In the compliance of SWM, these nations are lacking in various fronts of SWM. The major concern for the authorities dealing in MSW is the substantial upsurge in the urban population, causing stress on the environmental components like land, water and air. Some common issues that the majority of the developed nations are facing are discussed below:

Inadequate infrastructure, resources and technology to handle the MSW

- Lack of legislation and policies
- Absence of scientific MSW disposal facilities
- Insufficient knowledge of the basic principles of SWM
- Prevalence of practice of manual scavenging

The population is a prime factor in the fate of MSWM system of a city, as higher the population higher will be the quantum of the waste generated. With the limited resources in hand, most the ULBs in the developing countries fail to provide the services as per the need of the citizens. In past two decades, majority of the urban centres of the developing nations have experienced a sizeable increase in the population, in most of the cities migration of people from the rural area to urban areas is the major factor in the sharp surge in the population. This population explosion has stressed the limited resources of the urban authorities and has worsened the situation.

Inadequate resources and infrastructure is one of the major setbacks for the ULBs working in the developing nations and perpetuate them to practice sub-standard practices like using equipment beyond their economic life,

Solid waste management practices practiced in the capital cities of few developing nations are described below:-

4.7.1. Greater Buenos Aires, Argentina

Buenos Aires means 'Good airs' and was established way back in the year 1536 by Spanish solider named, Pedro de Mendoza. The city spans over 4,758 km^2 with a population of 15.5 million consisting of about 40% of the country's total population and is the financial, industrial, and commercial hub of the country (Quirós and Mehndiratta, 2015).

As far as the state of waste management of the city is concerned, the city has evolved with an inclusive waste management system to manage the humongous magnitude of the municipal solid waste and the state-owned organization named CEAMSE (Ecological Coordination Society of the State Metropolitan Area) is responsible for managing the MSW of the city. A significant portion of the population, i.e. 95.6% is covered with door to collection of waste with a frequency of twice a week. The city is also equipped with a robust waste transfer s system that is responsible for 76.3% of the total waste amounting 7,300 TPD. Out of the 23 zones of the city, 17 enjoy the existence of designated sanitary landfill, which accounts for 13,942 tonnes of waste generated by the households every day.

Holistic waste processing facilities were implemented by the authorities to crater the behemoth quantum of MSW generated like utilization of 2400

Table 2: MSWM system prevails in the developed countries

S.N.	City, Country	Population (millions)	Waste Generation Rate (TPD)	Presence of interventions				Waste Processing			Major Goal	References
				Policy	Institutional/ organization	Social	Technical	Rate of Recycling	Waste disposed of in landfills	Waste to energy		
1.	New York, USA	8	38,356	Yes	Yes	Yes	Yes	14%	76%	10%	Zero waste by 2030	(Cohen et al., 2015)
2.	Hong Kong	7.2	18,000	Yes	Yes	Yes	Yes	48%	36%	26%	Reduce the per capita disposal rate of municipal solid waste by 40% by 2022	(Cohen et al., 2015)
3.	Auckland, New Zealand	1.6	4.508	Yes	Yes	Yes	Yes	12%	20%	-	Zero Waste to landfill by 2040	(Auckland ' s Waste Assessment 2017, 2017)
4.	Sweden	10	33,170	Yes	Yes	Yes	Yes	23.8%	24%	50.2%	Not exceeding the waste generation and at the same time makes possible use of the waste produced.	(Avfall Sverige, 2018)

TPD capacity C&D Waste Processing Plant with a 90% recovery rate. For processing the organic fraction of waste, a mechanical-biological treatment plant of capacity 1000 TPD with a 50% recovery rate is in operation. The city is also equipped with a composting plant that transforms 10 TPD of the kitchen waste into soil conditioner for agriculture fields.

The authorities also encourage the concept of recycling, and about eight recycling plants were operating in the city that recycles 600 TPD of MSW and has successfully transformed the lives of 4,200 informal sanitary workers by inducting them into formal MSWM system.

However, still, 6,33,622 inhabitants suffer from the irregular waste collection service and lack door to door collection of waste. Also, 3,25,466 individuals live in the vicinity of open dumpsite possessing continuous risks of all the possible threats of the open dumpsite like groundwater contamination, nuisances caused by flies and odours, possible radioactive radiations, etc.

CEAMSE has played a commendable role in managing the wrecked state of the MSW prevailing in the city due to severe population stress on the city. Though the state authorities and citizens have to work in synergy by 100% implementation of the ideal waste management practices like segregation of waste at its source, reclamation of open dumpsites into sanitary landfills and works in the achievement of 'zero waste generation society'.

4.7.2 Mexico City, Mexico

Mexico City is the capital city of Mexico; in fact, it is the most populous of North America and is situated at an elevation of 2,250m above mean sea level. The city provides shelter to its 89,18,653 inhabitants and is one of the major economic hubs of Latin America and is the major contributor in the country's GDP as the city accounts 15.8% of the country's GDP.

In the compliance of MSWM the city is shifting its MSWM paradigm from conventional MSWM practices like the dumping of comingled waste into open dumpsites, poor collection of MSW from waste generators etc. to modern MSWM practices like sanitary landfills, adoption of integrated MSWM system etc. To channelize the MSW generated from various source to their respective treatment facility the AGU (Agencia de Gestión Urbana de Cuidad de México) has divided the city into 16 zones and established a waste transfer station in each zone.

On an average, the city generates out 12,893 TPD of MSW with a rate of 1.5 kg/capita/day against the world average of 1.2 kg/capita/day (Guibrunet *et al.*, 2016). The AGU has incorporated the concept of Public-Private Partnership in the MSWM and relies on the private firms on the collection, transportation, processing and disposal of the MSW generated in the city. However, informal waste pickers also play a significant role in the collection and recycling of the MSW and often termed as '*pepena*' which means 'to pick up' in the local language.

Thus the city experience a unique unison of all the three stakeholders namely AGU, private firms, and rag pickers and share responsibilities and resources to tackle the behemoth quantum of MSW to keep the city neat and free from the trash. But the state authorities are also lacking at the numerous fronts like policy implementation, personal protective and safety equipment especially in the case of informal waste pickers and research and market development for the small and medium scale recyclers.

Currently, the city is facing the land shortage crisis for setting up the new landfill site as nearly 97% of the MSW is dumped in the only landfill site namely "Bordo Poniente" spanning 370 hectares which have already attained its capacity in the year 2008. Subsequently, in 2011, AGU has taken a significant step in the strengthening the MSWM system and curbing the GHGs emissions of the city by releasing an order for closure of the main landfill site. Along with, the AGU had shown its commitment for setting up a landfill gas collection assemblage over the landfill to tap the high calorific value landfill gas and further utilize it in waste to energy expeditions.

The closure of the only landfill site of the city has raised several concerns in the handling of the magnanimous quantum of the MSW generated in the city and made the government come up with an integrated MSW management system.

4.7.3 Manila, Indonesia

The capital city of Indonesia, Manila is one of the world's oldest cities and densest populated city in Southeast Asia. It is situated on the eastern shore of Manila Bay. The city of Manila is divided into seventeen sectors namely Caloocan, Las Pinas, Makati, Malabon, Mandaluyong, Manila, Marikina, Muntinlupa, Navotas, Paranaque, Pasay, Pasig, Pateros, Quezon City, San Juan, Taguig and Valenzuela, conjointly generates 6,700 TPD. The MSWM practices are governed by 'Ecological Solid Waste Management Act, 2000' formally known as RA 9003). These rules came as the aftermath of the infamous Payatas dumpsite tragedy that has to lead to the death of 205 habitants and hundreds of them were gone missing. Responsibility of collection and disposal of MSW is held collectively by the ULBs and private contractors. Door to door collection of MSW is a prominent waste collection method for the majority of the households.

However, the rate of recycling or processing of collected is quite low as less than 10 per cent of the waste was recycled, and the majority of the collected waste is dumped into dumpsites. Also, illegal dumping and open burning of the waste is practiced as about 27 per cent of the waste is burned in open or subjected to illegal dumping. The collection of waste is carried out mainly through trucks. However, route management of deployed trucks is quite poor, and it is necessary to route optimization of the existing routes using remote sensing technique.

The poor state of the existing MSWM system could be cured by inducing incentives and motivation among the citizens to reduce waste generation. Public awareness campaigns could play a pivotal role in the movement. Also, revolutionary measures have been incorporated into the existing MSWM system through RA 9003. It mandates segregation of waste at its source and regularizes the waste processing procedure such as sorting of waste, composting, recycling, and establishment of e-waste collection centers.

4.8 Concluding remarks

Environmental protection laws and legislations had led to the inception of waste management regulations in the modern era. Based on the economic profile and standard of living, countries are termed as developed and developing nations. Both rich and developing countries have massive amounts of solid waste in the form of plastics, e-waste, residential garbage, biological waste, and hazardous waste. But due availability requisite of infrastructure, capital, and policies developing countries are managing the menace of solid waste quite well compare to the developing nations. Lack of policy and planning is found to be a common factor among the developing nations behind the poor state of solid waste management. Limited devoted capital funds, the sacristy of land, and the use of inferior technology lead to the high reliance on human resources in handling and management of solid waste. This makes the working conditions of employed personnel and the civic environment more vulnerable. In order to accomplish comprehensive solid waste management, developing nations should apply the community-focused integrated municipal solid waste management plan.

References

Ahmed, M.M., Shimada, K., 2019. energies The Effect of Renewable Energy Consumption on Sustainable Economic Development : Evidence 1–15.

Antoniou, G.P., De Feo, G., Fardin, F., Tamburrino, A., Khan, S., Tie, F., Reklaityte, I., Kanetaki, E., Zheng, X.Y., Mays, L.W., Angelakis, A.N., 2016. Evolution of toilets worldwide through the millennia, Sustainability (Switzerland). https://doi.org/10.3390/su8080779

Auckland ' s Waste Assessment 2017, 2017.

Balk, D., Montgomery, M.R., Engin, H., Lin, N., Major, E., Jones, B., 2019. Urbanization in India: Population and urban classification grids for 2011. Data 4, 1–16. https://doi.org/10.3390/data4010035

Bhat, R., Ahmad Dar, S., Ahmad Dar, D., Hamid Dar, G., 2018. Municipal Solid Waste Generation and current Scenario of its Management in India. Int. J. Adv. Res. Sci. Eng. 7, 419–431.

Bhattacharya, R.B.& A., 2017. Waste management by Ragpickers: the Tragic story behind India's informal sector. Banega Swachh India.

Bisarya, R.K., Puri, S., 2005. The Bhopal gas tragedy - A perspective. J. Loss Prev. Process Ind. 18, 209–212. https://doi.org/10.1016/j.jlp.2005.07.006

Central Government, 1995. The National Environment Tribunal Act , 1995 1995, 1–14.

Chandramouli, C., 2011. Provisional Population Totals, Office of the Registrar General & Census Commissioner, India Commissioner, India. New Delhi.

Chokhandre, P., Singh, S., Kashyap, G.C., 2017. Prevalence , predictors and economic burden of morbidities among waste-pickers of Mumbai , India : a cross-sectional study 1–8. https://doi.org/10.1186/s12995-017-0176-3

Cohen, B.S., Martinez, H., Schroder, A., 2015. Waste Management Practices in New York City , Hong Kong and Beijing 1–20.

CPCB, 2015. The National Action Plan for Municipal Solid Waste Management. New Delhi.

Dandapani, S., 2018. Unpaid and undervalued, how India's waste pickers fight apathy to keep our cities clean | The News Minute [WWW Document]. News Minute. URL https://www.thenewsminute.com/article/oppressed-and-unrecognised-life-waste-pickers-crucial-india-s-sanitation-72426

DeFrancis, M., 2011. The Economic Impacts of Inadequate Sanitation in India, Write Media.

Department of environment, 1986. Eprotect_Act_1986.Pdf.

Dutta, S., Upadhyay, V.P., Sridharan, U., 2006. Environmental management of industrial hazardous wastes in India. J. Environ. Sci. Eng. 48, 143–150.

Fewtrell, L., Colford, J.M., 2005. Water, sanitation and hygiene in developing countries: Interventions and diarrhoea-a review. Water Sci. Technol. 52, 133–142. https://doi.org/10.2166/wst.2005.0244

Forester, B.R.-I., 2017, undefined, 2017. Riddles in Schedules of Indian Wildlife (Protection) Act, 1972 in Need of Relook. Indianforester.Co.in 143, 985–992.

Ganguly, S.C., 2008. India's national sanitation and hygiene programme : From experience to policy West Bengal and Maharashtra models provide keys to success, in: Rural Sanitation. pp. 126–147.

Ghosh, A., Cairncross, S., 2014. The uneven progress of sanitation in India. J. Water Sanit. Hyg. Dev. 4, 15–22. https://doi.org/10.2166/washdev.2013.185

Guerrero, L.A., Maas, G., Hogland, W., 2013. Solid waste management challenges for cities in developing countries. Waste Manag. 33, 220–232. https://doi.org/10.1016/j.wasman.2012.09.008

Guibrunet, L., Sanzana, M., Castán, V., 2016. Geoforum Flows , system boundaries and the politics of urban metabolism : Waste management in Mexico City and Santiago de Chile. Geoforum. https://doi.org/10.1016/j.geoforum.2016.10.011

Handl, G., Deutsch, E., Law, I., 2012. Historical Archives - Introductory Note - Declaration of the United Nations Conference on the Human Environment (Stockholm Declaration), 1972 and the Rio Declaration on Environment and Development, 1992 - English 1–11.

Herrera, V., 2019. Reconciling global aspirations and local realities: Challenges facing the Sustainable Development Goals for water and sanitation. World Dev. 118, 106–117. https://doi.org/10.1016/j.worlddev.2019.02.009

Jain, S., 1984. Water (Prevention and Control of Pollution) Act 1974: Basic Legal Issues.

Joshi, R., Ahmed, S., 2016. Status and challenges of municipal solid waste management in India: A review. Cogent Environ. Sci. 2, 1–18. https://doi.org/10.1080/23311843.2016.1139434

Kershner, E., 2020. How Many Countries Are There In The World? Https://Www.Worldatlas.Com.

Khandelwal, H., Dhar, H., Thalla, A.K., Kumar, S., 2019. Application of life cycle assessment in municipal solid waste management: A worldwide critical review. J. Clean. Prod. 209, 630–654. https://doi.org/10.1016/j.jclepro.2018.10.233

Kumar, S., Bhattacharyya, J.K., Vaidya, A.N., Chakrabarti, T., Devotta, S., Akolkar, A.B., 2009. Assessment of the status of municipal solid waste management in metro cities, state capitals, class I cities, and class II towns in India: An insight. Waste Manag. 29, 883–895. https://doi.org/10.1016/j.wasman.2008.04.011

Kumar, S., Smith, S.R., Fowler, G., Velis, C., Kumar, S.J., Arya, S., Rena, Kumar, R., Cheeseman, C., 2017. Challenges and opportunities associated with waste management in India. R. Soc. Open Sci. 4. https://doi.org/10.1098/rsos.160764

Management, S.W., 2018. Swedish Waste Management.

Mandpe, A., Kumar, S., Lakshmikanthan, P., Hettiarachchi, H., 2019. Mining for recovery as an option for dumpsite rehabilitation: Case study from Nagpur, India. J. Environ. Eng. Sci. 15, 52–60. https://doi.org/10.1680/jenes.19.00021

Mehrotra, S., Ghosh, D., 2013. Addressing the world's worst sanitation problem : a programme re-design to use not just build toilets.

MOEF, 2016. Hazardous and Other Wastes - Ministry of Environment and Forests. The gazzate of India 1981, 1–68.

Mohapatra, G., 2019. Projected Behavioural Change in Swachh Bharat Mission: A Public Policy Perspective. Indian J. Public Adm. 65, 451–474. https://doi.org/10.1177/0019556119863856

MoHUA, 2018. Government of India Ministry of Housing And Urban Affairs Lok Sabha Unstarred Question No. 5782.

Nambiar, M., 2001. Making the Gram Sabha Work. Econ. Polit. Wkly. 36, 3114–3117.

NASA, 2016. Fire Burns in Mumbai Landfill [WWW Document]. URL https://earthobservatory.nasa.gov/images/87429/fire-burns-in-mumbai-landfill

Of, A.C.T.N.O., 1992. The Public Liability Insurance Act, 1991 An Act to provide for public liability insurance for the purpose of providing immediate relief to the persons affected by accident occurring while handling any hazardous substance and for matters connected therewi 3, 1–8.

Paliya, S., Mandpe, A., Kumar, S., Kumar, M.S., 2019. Enhanced nodulation and higher germination using sludge ash as a carrier for biofertilizer production. J. Environ. Manage. 250, 109523. doi:10.1016/j.jenvman.2019.109523

Patel, M.L., Jain, R., Saxena, A., 2010. Assessment of the Municipal Solid Waste & Status of Implementation of Municipal Solid Waste (Management & Handling), Rules, 2000 in the State of Madhya Pradesh, 2008 - A case study. Waste Manag. Res. 29, 558–562. https://doi.org/10.1177/0734242X10372662

Planning, F., 2017. Bodies in Poverty liI.

Quirós, T.P., Mehndiratta, S.R., 2015. Accessibility analysis of growth patterns in Buenos Aires, Argentina: Density, employment, and spatial form. Transp. Res. Rec. 2512, 101–109. https://doi.org/10.3141/2512-12

Rules, H., Waste, B., Methods, T., 2019. Bio-Medical Waste Management Practices in India – 6, 364–372.

Sambodhi Research and Communications, Wankhade, K., Sambodhi Research and Communications, 2015. Access to toilets and the safety, convenience and self-respect of women in rural India. Environ. Urban. 27, 20.

Sampath, V., 2019. Gandhi valued sanitation more than Independence [WWW Document]. The

Print. URL https://theprint.in/opinion/gandhi-valued-sanitation-more-than-independence-modis-swachh-bharat-is-fulfilling-his-dream/220306/ (accessed 11.18.21).

Sarkar, R., 2019. Urbanization in India Before and After the Economic Reforms: What Does the Census Data Reveal? J. Asian Afr. Stud. 54, 1213–1226. https://doi.org/10.1177/0021909619865581

Sastry, N.S., 1999. Drinking Water , sanitation and hygiene in India, National Sample Survey Organisation Department Of Statistics Government Of India.

Singh, P., Sharma, V.P., 2016. Integrated Plastic Waste Management: Environmental and Improved Health Approaches. Procedia Environ. Sci. 35, 692–700. https://doi.org/10.1016/j.proenv.2016.07.068

Singh, S.K., Chokhandre, P., Salve, P.S., Rajak, R., 2021. Open dumping site and health risks to proximate communities in Mumbai, India: A cross-sectional case-comparison study. Clin. Epidemiol. Glob. Heal. 9, 34–40. https://doi.org/10.1016/j.cegh.2020.06.008

Sundararajan, V., 2013. Mangalyaan - Overview and technical architecture of India's first interplanetary mission to mars. AIAA Sp. 2013 Conf. Expo. 1–9. https://doi.org/10.2514/6.2013-5503

The Hindu, 2017. Fire breaks out at Bhalswa landfill. The Hindu.

UNICEF, 2011. CensusInfo India 2011, Census of India 2011.

United Nations, 2018. World Urbanization Prospects, 2019th ed, Demographic Research. New York.

United Nations, 2016. Functions and powers of the General Assembly. Website Gen. Assem.

Vishwakarma, D., 2016. Swachh Bharat Abhiyan Clean India Abhiyan. Int. Res. J. Manag. IT Soc. Sci. 3, 48–52. https://doi.org/10.21744/irjmis.v3i3.94

Wang, Q., Su, M., Li, R., 2018. Toward to economic growth without emission growth: The role of urbanization and industrialization in China and India. J. Clean. Prod. 205, 499–511. https://doi.org/10.1016/j.jclepro.2018.09.034

World Health Organization, 2005. mdg Health and Millennium Development Goals. World Heal. Organ. 13.

Zhang, C., Xu, T., Feng, H., Chen, S., 2019. Greenhouse gas emissions from landfills: A review and bibliometric analysis. Sustain. 11, 1–15. https://doi.org/10.3390/su11082282

5

Challenges Associated with Waste Management in Developing Countries

Hardeep Rai Sharma[a]* and Nisha Rani[b]

aInstitute of Environmental Studies, Kurukshetra University Kurukshetra-136119, Haryana, India
bDepartment of BioSciences, Himachal Pradesh University, Shimla-171005 Himachal Pradesh, India

ABSTRACT

Different types of wastes are generating from different settings and different processes. Waste disposal is a big issue in many world countries. Besides having rules and regulations, allotted budget, available infrastructure, many waste management authorities face problem during proper waste disposal and as a result, leads to environmental and public health problems, including community discomfort. As a failure, scattered wastes are commonly seen in many public places, behind the bushes, empty plots inviting all sorts of disease vectors, nuisance and odour. Various waste management and disposal techniques are available and are used depending upon various factors like types, amount, characteristics of wastes, and availability of techniques and funds. Six fundamental aspects of waste management, starting from the generation of wastes and will continue until their final disposal. Some of these techniques generate wealth in the form of energy, compost and recyclable materials. Different types of rules and regulations are also present to guide and check proper solid waste management. Although all these sustainable wastes management become a challenge for the common man, policymakers, governments and scientists. The present chapter will discuss the various engineering, scientific and organisational challenges of waste management and a possible way forward.

Keywords: Waste, management aspects, landfilling, incineration, composting, energy generation, challenges, recycling, reuse.

Definitions: Before starting the chapter, it is better to understand the basic definitions of the terms commonly used in this chapter.

Waste: Waste is any material or object which is discarded after its primary use by its holder. A holder means the person who produces and keeps the waste.

Solid waste: Wastes which are generally solid and generated from various human and animals activities.

Hazardous wastes: According to U.S. Resource Conservation and Recovery Act of 1976, any waste that is hazardous and toxic if it causes or considerably contribute to rise in mortality or serious irreversible, or incapacitating reversible illness; or pose a significant present or potential hazard to the environment and human health when not treated, stored, transported, disposed or managed properly. Hazardous waste has the properties of toxicity, reactivity and corrosiveness.

Household waste: Originated from residential homes, private domestic accommodations, campsites, private garages, universities/schools, hospitals premises, royal palaces, prisons and meeting halls.

Industrial waste: The waste produced from any individual organisations, and industries, *e.g.* waste produces from food processing, textiles, sugar, furniture, plastic, steel, chemical industries *etc.*

Commercial waste: Originates from sites used mainly for a business or trade or recreation or the purpose of sport, *e.g.* waste from hotels offices, shops, market, local authorities, fairs *etc.*

Clinical waste: It includes any waste that arises from medical, dental, nursing, veterinary, pharmaceutical, treatment, teaching and research activities, which comprises mainly or partly of animal or human origin tissue, blood or other body fluids, secretions, medicines or other pharmaceutical products, dressings, needles, syringes, blood transfusion, packing associated with the waste. This type of waste is also known as hospital waste or infectious waste.

Municipal solid wastes: Waste collected and disposed of by or on behalf of a local authority commonly by municipalities. In general municipal solid wastes (MSW) consist of mainly commercial and household wastes. The waste collected from local authority *i.e.* construction and demolition waste; gully sweepings, street sweepings as well as waste collected from civic amenities waste collection sites also included in MSW.

Inert wastes: Waste which does not undergo any significant physical, chemical or biological changes are considered as inert waste. These wastes do not physically or chemically react, burn, dissolve, biodegrade or adversely affect other matter with which comes their contact in a manner that may cause pollution or harm to environment and human health. Examples of inert wastes include uncontaminated earth and excavation waste, *i.e.* bricks, stones, concrete, building sand and gravel, ceramic materials and slates.

Controlled waste: Wastes which can be controlled, *e.g.* industrial, commercial and household waste.

Uncontrolled waste: Waste which cannot be controlled *i.e.* agricultural, mine and quarry waste. The uncontrolled waste is produced in large quantities and are subject to separate legislation.

5.1 Introduction

According to the World Bank (2012) forecast, the global municipal solid waste generation will twofold by 2025 (2012 considered as a reference year). As per Waste Atlas (2017) India is the third-largest generator (volume basis) of municipal solid waste. Different types of wastes are generating from different settings and different processes. Various management and disposal techniques are available and are used depending upon various factors like types, amount and characteristics of wastes. Six fundamental aspects of waste management, starting from the generation of wastes and will continue until their final disposal. Some of these techniques generate wealth in the form of energy, compost and recyclable materials. Different types of rules and regulations are also present to guide and check proper solid waste management. Even though having all this, sustainable wastes management become a challenge for the common man, policymakers, and scientist. Inadequate waste infrastructure, informal sector and waste dumping along with people attitude towards waste are the main issues associated with solid waste management in India (Kumar et al., 2017). According to ERM (2004) in addition to poor administrative management practices, an allotment of the budget is constrained in the solid waste sector and has been a focus of increasing concern.

5.1.1 Classification of Wastes

Before discussing the challenges, it's better first to understand the types of wastes as proper management will depend on it. Based on sources of generation, solid waste can be categorised as non-process waste, *i.e.* residential, commercial, institutional, industrial; construction and demolition waste, municipal solid waste and municipal services (excluding treatment facilities), agricultural, and industrial wastes (Tchobanoglous et al.,1993, Biswas et al., 2010; Dhar et al., 2017; Kumar et al., 2017, 2009; Wainaina et al., 2020which are facing a serious pollution problem due to the generation of huge quantities of solid waste. This paper presents an assessment of the existing situation of municipal solid waste management (MSWM). Solid waste can be classified into food waste (garbage), rubbish (combustible and non-combustible), street waste, bulky waste, ashes and residues, industrial waste and sludge, dead animals, construction and demolition waste, hazardous and non-hazardous wastes and horticulture wastes (Bhide and Sundaresan,1993; Paliya et al., 2019). Based on waste to energy (WtE) applications, MSW can be classified into inert fraction/inorganics/bulky materials; compostable organics and refuse-derived fuel (RDF) (woody wastes, paper, plastics, and textiles). Based upon the source, properties and types, solid waste can be categorised into different categories (Fig. 1).

Source Based

- Industrial Waste
- Miuucipal Solid waste
- Commercial/ Institutional waste
- Biomedical Waste
- Agricultural Waste

Property based

- Organic / Biodegradable Waste
- Inorganic/Non- biodegiadable waste
- Hazardous/non- hazaidous waste

Type based

- Rubbish
- Bulk waste
- Garbage
- Ashes
- Street waste
- Deadammals
- Construction and demolish waste
- Sewage/sludge
- Plastic

Fig. 1: Solid waste classification based upon *Source*, Properties of waste and Type of waste

5.1.2 Functional Elements of a Waste Management System

There are six main functional elements described below

1. ***Waste generation***: In includes activities which identified materials as no longer value and are gathered together for disposal, *e.g.* chewing gum/toffee/chocolate wrapping is usually considered to be of little importance to the holder after their respective consumption. Waste generation

activity presently is not very controllable. To limit the quantity of waste generated, the reduction measure can be taken at source. According to Solid Waste Management (SWM) Rules, 2016 it is the responsibility of waste generator to segregate wastes in separate categories, and store in suitable dustbins before it is handed over to authorised waste collectors.

2. ***Waste handling: segregation, storage and processing at the source***: Handling of waste is very important in the entire management system as it involves all the activities associated until it is placed in storage container and further transportation of loaded container to the collection point. Separation is a significant stage in the handling and storage of waste at source. To recover recyclables like plastics, glass, metals *etc.* it is best to separate them at the source. Onsite storage is very crucial because of aesthetic considerations and public health concern. The cost of providing storage usually is bear by the homeowner or apartment owner or management of commercial or industrial properties. The solid waste is further process by yard composting and compaction.

3. ***Collection***: Solid and recyclable materials collected and gathered to be transported to the locations identified for emptizing collection vehicles.

4. ***Segregation, processing and transformation of solid waste***: waste is taken away from its generation points where further processing and transformations take place. Processing and transformation of waste includes the separation of bulky items and other waste components manually, through screens, using magnets for separating iron. Separation of source-separated waste and commingled waste mainly occurred at transfer stations, materials recovery facilities, disposal sites and combustion facilities. Size is reduced by shredding and volume reduction is done by combustion and compaction.

5. ***Transfer and transport***: It is carried out through the following steps:

    a) The wastes are transferred from the smaller collection vehicle to the bigger roader transport equipment.

    b) It is then subsequently transported to a processing or disposal site over long distances. Although the motor vehicle is widespread in use to transport waste, rail cars and barges (flat bottom large boats) are also used to transport wastes. One of the largest rail haul operations currently in use is used to transport wastes from the city of Seattle, Washington (WA) to the Columbia Ridge landfill located approximately 300 miles away in Gilliam, OR, USA. Operationally 25 to 28 tons of waste is compacted into a 40*ft*, and sealed shipping container mounted on a trailer hulk. In some parts in England, river barges are used to carrying wastes. The hydraulic as well as pneumatic transport systems, use both low pressure air and vacuum conduit to carry solid wastes. The most common application of this system can be seen in the transport of waste from high–density commercial areas or apartments to a central site meant for processing or loading into transport vehicles. The most extensive pneumatic system was installed at the Walt Disney World amusement park in Orlando, Florida in America.

6. ***Disposal**:* It is the last and most important component in the solid waste management system. In several parts of the world, even today the ultimate fate of all solid waste is landfilling or land spreading. Several methods of solid waste disposal are recycling, composting, controlled tipping, burial, hog feeding, regular open dumping, incineration/ combustion, sanitary land-filing, grinding and discharge into sewers and in the modern sanitary landfilling, the solid waste is disposed on land or within the earth's mantle using an engineered facility in such a way that it does not create any nuisances or hazards to public health, such as contamination of groundwater, breeding of insect and rats. Currently, sanitary landfilling is most opted method followed by incineration to dispose of the solid municipal waste (Mandpe et al., 2019a).

Waste disposal is a big issue in many world countries. Besides having rules and regulations, allotted budget, available infrastructure, many waste management authorities face problem during proper waste disposal and as a result, leads to environmental and public health problems, including community discomfort. As a failure, scattered wastes are commonly seen in many public places, behind the bushes, empty plots inviting all sorts of disease vectors, nuisance and odour. Various challenges associated with waste management in developing countries can be categorised into engineering, scientific and organisational one.

5.2 Engineering Challenges

5.2.1 Scientific Collection of Waste

Collection includes not only to gather or to pick up of solid wastes from the different sites but also carrying of these wastes to the location where the content of the collection vehicles is emptied. The unloading of the collection vehicle is also considered part of the collection operation. In residential areas, the collection is done by management or by solid waste disposal companies. In industries and agricultural areas, it is the responsibility of the owner. Collection of separated and commingled solid waste from an urban area is complex and challenging because:

- Generation of waste takes place every residential house, apartments and commercial including small industries as well as in the parks, streets, and even empty areas.
- Suburban areas have mushroom-like growth all over the country.
- Pattern of waste generation became more diffused with an increased quantity of waste, making waste collection more difficult.
- High costs of fuel and labour, *e.g.* generally 50-70% of the money used in the collection of solid waste out of total money used for solid waste management.

The most common method for collection of residential waste is manual one where loaded containers are lifted directly and empty to the collection vehicle. In some cases, either loaded containers are rolled on their rim or small lift is used to take them towards collection vehicles.

5.2.1.1 Types of Collection Services

The collection of waste vary depending upon the type and built of houses. The most commonly used are as follows:

From low-rise detached houses: In this type of collection system following services are included:

Curb service, the person living in the house place the container of waste to be emptied at the curb on collection schedule and return the empty container to its position until the next collection.

Alley, it is the part of the basic layout (path, area covered or surrounded by plants *etc.*) of a city or a given inhabited area.

Setout-setback service, boxes or dustbins are set out from the homeowner's property and setback after being emptied by additional crews with collection vehicles.

Similarly, in setout service the empty container is returned to its original storage location by the owner itself.

For **curb collection**, sometimes satellite (small) vehicles are used. These satellite vehicles then emptied into big containers. These satellite vehicles are used in congested areas where big vehicles cannot enter to collect waste.

From low and medium-rise apartments: Garbage is collected commonly through curbside collection service where maintenance staff is incharge for carrying the containers mechanically or manually to curbside collection location.

From high-rise apartments: Large container collection system is used commonly.

From commercial and small industrial settings: Waste are collected through manual as well as mechanical means. To avoid traffic congestion during the day, early morning and late evening hours are chosen to collect solid waste in many large cities. Two-three collectors with one driver in a crew collect the wastes. Plastic bags, cardboard boxes, disposable containers are used for manual collection in curb type collection system. Movable containers are also used to avoid the problem of space for storing containers.

5.2.1.2 Collection of Waste Separated at the Source

Curbside collection method with specially designed conventional collection vehicles are used most widely in this type of collection system. Sometimes this is done by charitable organisations in which delivery is done by homeowners to drop-off and buys back centres.

The curbside program is more popular and has higher participation than for drop-off programs as residents and businessman do not have to transport the recyclable any further. Some program requires inhabitants to separate and store different wastes *e.g.* glass, plastic, newspaper and metals which are later collected separately. Sometimes one container is used to collect all recyclable materials and at another time different containers are used to collect paper and heavy recyclable like glass, tin wastes *etc.* As per SWM Rules, 2016, three colours for dustbins are – green for storage of biodegradable wastes, white for recyclable wastes and black for storage of other wastes.

5.2.1.3 Types of Collection Systems

Collection systems can be categorized into two types according to the method of operation as discussed below:

Hauled containers systems (HCS): The wastes are collected in relatively large containers that hauled to disposal site and after being emptied returned to the original location. These HCS are best suitable for the sites where waste is generated in large quantities and needs frequent collection. Different sizes and shapes of hauled containers are available as collection with large containers reduces time for handling and hideous accumulations of waste.

In this system (HCS) each container picked up and requires only driver and one truck to complete the collection cycle through a round trip to the disposal site. As containers are filled manually, loading aids such as ramps and platforms are provided to use huge containers that often lead to low-volume utilisation. The hauled container systems can be of: i) Hoist truck type where the large capacity of mechanically loaded collection vehicles are employed, ii) Tilt-frame container system using tilt frame-loaded vehicles and large containers, often called a drop or debris boxes, and iii) Trash-trailer generally suitable for bulky debris, such as metal scrap, sand, timber, and are similar to tilt-frame type.

Stationary container systems: In this type, the collection containers remain stationary and are not carried over to the disposal site for emptying. Further on the basis of the containers loading, these are of two types *i.e.* mechanical and manual.

As per SWM rules, 2016 the local authorities and village councils (gram panchayats)are responsible for the arrangement of door to door collection of separated solid waste from all residences including slums and informal settlements, institutional, commercial and other non-residential premises. However, from high rise apartments, big commercial complexes, shopping malls, residential complexes waste may be collected from the entrance or any other designated location. In Indian cities, some of the problems faced by the urban local bodies during MSW waste collections are low collection coverage and irregular collection services, irregular and inefficient transportation facilities, and non-cooperation from household, trade and commerce in waste management practices (Vikas, 2017). To manage the solid waste in rural areas in the Haryana state, a project named *Kachre Se Kamai* (Wealth from Waste) was initiated under the Nirmal Bharat Abhiyan of Ministry of Drinking Water and Sanitation (MDWS) in 2012. Under this project about 1336 solid waste management projects were sanctioned in 22 districts of Haryana from the year 2014 to till date out of which about 50 % were completed, 11 % were in progress of construction, and about 39 % were yet not started. Many of these projects faced problems of waste segregation, collection, management, human resources and economic issues (Kumar and Sharma, 2019).

5.2.2 Transportation, Segregation, and Disposal of Waste

For maintaining cleanliness in an area door to door collection, regular sweeping, removal of garbage and covered transportation are the required criteria. The waste collected in community dustbins is transferred to transport vehicles for transport to the processing or disposal site. In big cities where disposal facilities are quite far, the collected wastes are conveyed to a transfer station from where another set of vehicles transport it further. Transportation of waste is done by using hand cart, tractor-trolley, trailer, tipper truck, *etc.* As observed in Kurukshetra city of Haryana, the transportation trucks are covered to avoid waste spillage and waste paper/plastic blowing.

Segregation is the primary separation of diverse categories of waste produced at source and hence, reducing the risks and the cost of handling and disposal.

A primary accepted important truth of source reduction is to avoid mixing wastes. Another basic rule-don't make it a liquid if it is dry. In many cities of developed countries, the garbage collected is transferred to the material recovery stations or integrated processing and transfer facility centres, wherein the compostable organic material is separated and conveyed to the onsite composting facility (Kumar et al., 2018). After separation, the remaining waste is processed into refuse-derived fuel (RDF), which is used as a feedstock for waste-to-biofuels plants. In many societies and places, different colour dustbins are provided by the municipalities to promote waste segregation at source, also known as recovery.

Further attempts have been made to promote waste segregation at collection level as collection vehicles are partitioned into two different compartments, one for dry and another for wet waste. However, these types of waste separation practices need people awareness and involvement. Segregation is a crucial step in composting and vermin-composting process where segregated biologically degradable wastes are converted to produce manure. Separation is also crucial before waste incineration, where the combustible fraction of waste is burned to process MSW (Manyapu et al., 2019).

Gitanjali Industries, a privately owned industry works in co-ordination with the Municipal Corporation of Greater Mumbai in Mumbai, is an excellent example of a public-private partnership for dry waste recycling. The industry has employed 350 waste pickers, to collect waste of different kind like mineral water bottles, plastic bags, e–wastes, scrap iron *etc.* The collected wastes are transported to the industrial unit and then sorted by twenty sorters. The industry has its marketing outlets in 10 cities, namely Bengaluru, Belgaum, Bhubaneshwar, Gurugram, Goa, Mumbai, Jammu, Ranchi and Sambalpur. The Gitanjali industry through its workers also maintains community bins, from where dry garbage is collected by rag pickers and has also adopted twenty-five gutters in the area surrounding its unit (Palnitkar, 2011).

5.2.2.1 Disposal of Waste

There are various methods of waste disposal. The choice of method depends on available resources, technology, types of wastes, availability of space *etc.* Each method has its advantages and disadvantages. However, while choosing the disposal method, one must consider the feasibility and possible solutions to overcome the environmental impacts if any related to that method. At this moment, the typical waste disposal methods are discussed under basic methods and practical techno-economical solutions in brief.

Open Dumping: Non-combustible rubbish, street sweepings, dirt and ashes are disposed by open dumping. Mixed solid wastes and garbage and any other waste that create odour, nuisance and hazardous for health are not suitable for open dumping. Open dumping of such waste can be a source of shelter and food for rats, flies and other insects. As observed in many cases on uncontrolled burning, it will produce smoke and release air pollutants. Many researchers reported scattered wastes, uncontrolled burning, blowing of waste plastics and papers by wind, odour from the waste piles, presence of cows, pigs, dogs and scavengers, lack of fencing and accessible roads overall creates an un-aesthetic view of the disposal sites (Sharma et al., 2011,2013, Tyagi et al., 2014). If no option is possible, the site for of open dumping should be cautiously selected so that there will be no complaints of inhabitants and minimum chance of water contamination.

Controlled Tipping/burial: Unsystematic dumping of waste creates favourable environment for insects to breed, food and shelter for rodents *etc.* Therefore, such problems can be avoided by disposing garbage under proper sanitary conditions. Among various methods, one simpler and cheaper methods is burying waste under controlled conditions. Under this a shallow trench is digged out, the waste is laid down in a systematic way, compacted mechanically as well as manually and at the end of each day the trench is filled by available soil or ash.This process is repeated every day systematically at suitable sites. This method can be used at locations where appropriate land is not a constrain and freely to be used for such practices. ***Animal Feeding***: The feeding of garbage, especially waste food, kitchen wastes to domestic animals and hogs or pigs has been practised for several years in various parts of India. Garbage feeding can be profitable if handled adequately by waste feeders by daily collecting it in clean containers.

Grinding and discharge into sewer lines: This method deal of disposing garbage or kitchen waste into sewers: There is household grinders which grind the kitchen wastes (except bones) for their easy disposal and transport with wastewater. But, the disadvantages associated with this is increased level of BOD as well as solids in sewage treatment plants. The municipalities install the central grinding system at convenient point along with sewer system. These are used to avoid blockage of sewers, but care should be taken from accumulated garbage, increases in >50% of suspended solids and more consumption of water.

Other methods: Methods like dumping of waste into drains, oceans, burning openly are also practised at many places; however, ocean dumping of wastes is already banned.

5.2.3 Practical Techno-economical Solutions

An environment-friendly, economically affordable, and socially acceptable, waste management system started from the source generation up to their final disposal, can be termed as "Sustainable". According to Indian Planning Commission report of 2014, if unused MSW can be managed appropriately it will generate power of about 439 MW, produce 1.3 million cubic metres/ day biogas or 72 megawatts of electricity from biogas and compost about 5.4 million metric tonnes annually (Planning Commission,2014) Waste-to-Energy (WtE) technologies like sanitary landfilling, incineration, plasma arc gasification, composting and vermicomposting and biogas production are the methods which if implemented properly can be sustainable and convert wealth and energy from the waste. By using effective pollution control technologies with these methods, environmental impacts can be reduced to much extent. Many countries, including India, has adopted and installed various WtE technologies and thereby generating electricity and wealth from wastes.

5.2.3.1 Sanitary Land Filling

The most commonly used method for waste disposal nowadays is landfilling or land disposal. Solid waste residues which cannot be recycled and left out even after the recovery of recyclable materials or energy can be disposed of by this method. However, in many cities of India where segregation and recycling opportunities are not available the municipal and other generated wastes are dumped by this method. This disposal method is a safe and reliable long-term option of solid waste residues disposal. Under the Capacity Building Program on Implementation of SWM Rules (2016) land requirement for landfill and waste processing is specified based on population. For a population of 1-5 lakh, 15 acres of land is mentioned for landfilling. Similarly for 5-10 lakh and >10 lakhs, 12 acres and 8 acres, respectively is specified along with one acre for waste processing in all cases. The district collectors may identify and choose large segments of wasteland for sanitary disposal away from habitation. This wasteland should meet the requirement for 20-25 years' of the cities/towns coming within 50 km radius from the landmarked and establish a mechanism of its professional management on a cost-sharing basis. Further SWM Rules-2016 avoided the construction of landfill on the hills. For the hilly areas the inert and the residual waste from processing unit shall be collected and taken to a transfer station at a suitable enclosed location. The open areas down the hill within 25 km should be identified and selected for setting the sanitary land fill site where residual waste form transfer station can be disposed of.

Preparation of the site for the landfilling: For preparing the site for landfilling first the existing site drainage is modified to route any runoff away from the intended landfill area. There must be the provision of access roads and weighing facilities and installation of fences so that the sites can be accessed easily. The next step in the development of a landfill is the excavation and preparation of landfill bottom and subsurface sides. Modern landfills typically are constructed in sections where only a small part of the unprotected landfill surface is exposed to precipitation. Excavated material from the landfill can be stockpiled on unexcavated soil near the active area in such a manner that it will minimise the problem of precipitation. Where the entire bottom of the landfill is lined at once, provision must be made to remove rainfall and stormwater runoff from the portion of the landfill that is not being used. To minimise costs, it is desirable to obtain cover materials from the landfill site whenever possible. Vadose Zone (the zone between the ground surface and permanent groundwater) and groundwater monitoring equipment is installed before the landfill liner is laid down. A low-permeability liner is installed and bottom of landfill is shaped to provide drainage of leachate. The facility for leachate collection and extraction are placed within or on top of the liner. Typically, the liner extends up the excavated walls of the landfill. Horizontal gas recovery trenches may be installed at the bottom of the landfill, particularly if emissions of volatile organic compounds (VOCs) from the newly placed waste are expected to be a problem. To minimise the release of VOC's a vacuum is applied and the air is drawn through the completed portions of the landfill. The gas that is removed must be burned under controlled conditions to destroy the VOC's. Before the fill operation begins, a soil beam is constructed at the downward side of the planned fill area. The beam serves as a windbreak to control blowing materials. The compaction of sanitary landfills must be ensured to prevent fire, reduce settlement problem, provides a hard ground for moving vehicles and future use in transport, eliminate odour, kill flies and to conserve trench space. Further a separate trench or pit is constructed for removal of large quantities of untraceable materials, spoiled foodstuff and small dead animals. Usually, the decomposition rate of waste is a slow process, in some areas, it may take about 10-15 years and the same site cannot be used for many years.

5.2.3.2 Incineration/Thermal Treatment

In this treatment, the combustible components of waste are burned at a very high temperature of 900-1100 °C in the presence of oxygen. Incineration is referred to as the best alternative to the disposal of waste. Biodegradable wastes are separated from the commingled waste before its incineration as only combustible materials are incinerated. The solid wastes are converted into gaseous, liquid and solid products with release of heat energy. The thermal processing of solid waste is an essential element in many integrated waste management systems used both for volume reduction as well as for energy recovery. It is the second method/option for solid waste disposal after landfilling. The thermal processing system can be categorised into mass burn incineration, gasification, pyrolysis and plasma. Based on air requirement incineration can be done with different types a) stoichiometric combustion with precisely the same amount of oxygen (or air) needed for complete combustion; b) excess air combustion with excess of stoichiometric oxygen; c) gasification is the partial combustion under sub-stoichiometric conditions to generate a combustible gas containing hydrogen, CO and gaseous hydrocarbons; d) pyrolysis in which thermal processing of waste is done in the absence of oxygen.

Generally, incinerators are classified as closed and open system. Under closed system a special chamber is designed with various parts to facilitates incineration while in open system the chamber is open to air. The incinerator chamber requires a chimney with proper height to provide a good flow of air through the combustion chamber. Based upon burning conditions, design aspects and amount of waste burned at a time, there are different types of incineration system.

Mass burn incinerator: It is widely used throughout the world to treat and disposed MSW. It's better first to assess the 'fuel' properties, *i.e.* ash, moisture content, volatile content and elemental analysis of the generated for the effective incineration and to control air emissions. From the design point of view, mass-burn incinerator has five components like furnace; bunker and feeding system; waste delivery, heat recovery, and pollution control. Incineration of waste is done in three stages of drying and incineration, burning of volatiles as well as soot particles and incineration of the solid carbonaceous residue. Initially, the moisture of waste is driven off at 50-1000C. After drying the wastes is exposed to pyrolysis and thermal decomposition of any organic material such as food waste, plastics, paper, and textile *etc* (Pratap et al., 2021). As a result, the volatile matter, vapours and combustible gases are generated. The temperature of furnace is maintained between 750 and 10000C as temperature higher than 12000C can leads to ash fusion and building of slag on refractory material.

Fluidised Bed Incinerators: These are designed for a large range of liquid and gaseous wastes that includes sewage sludge, municipal solid and hazardous wastes. It consists of a refractory lined chamber containing a bed of sand particles. The primary combustion air is entered from below in the chamber, which makes the sand particles fluidised. The fluidised bed reactor promotes rapid heating to ignition temperature, dispersion of incoming waste and sufficient residence time in the furnace for complete incineration. The secondary function includes the uniform heating of excess air, excellent heat transfer for heat exchange surfaces within the bed to adsorb pollutants like the addition of lime (CaO) to reduce SO_2 emissions. This type of reactor significantly increases the burning rate of waste, and enhanced the rate of new char formation. The char oxidation also enhanced as the charred surface of the burning waste is continuously abraded by the bed material.

Starved Air Incinerators / Pyrolytic Incinerators: This type of incinerators are widely used for incineration of industrial and clinical wastes. Pyrolytic incinerators also known as starved and are smaller in size with a capacity ranging from 5 to 100 tonnes per day. These incinerators consist of combustion in two chambers *i.e.* pyrolytic and combustion. In the pyrolytic chamber, the waste is chemically decomposed by the action of heat in the inert atmosphere. The two-stage combustor relies on semi-pyrolysis, where sub-stoichiometric combustion of the waste produce heat for the thermal decomposition or gasification of the waste. Due to insufficient air, high percentage of incomplete combustion products are formed which passes to the second stage. At first stage due to heat generated during pyrolytic process the temperature of the gases leaving this section rises to 700-800 ^{0}C. These gases then passed to the secondary section *i.e.* combustion chamber with excess secondary air, where approximately 200% stoichiometric conditions to give a temperature of 1000-1200 ^{0}C. In the combustion chamber complete burning of gases (hydrogen, carbon monoxide and hydrocarbons) has taken place and formation of NOx is inhibited. Dioxins and PCBs present in the secondary gases are destroyed in secondary chamber. The long residence time and high temperature of over 10000C in this chamber facilitates this reaction. The ash is finally released at the rear of the incinerator into a water trough. The low combustion air flow results in low entrainment of particulate matter in the flue gases, which also decreases the other particulate – borne pollutants such as dioxins, furans and heavy metals. The two-stage processes ensure relatively low gas velocity, because of which most particulate matter retained in the first chamber.

Rotary kiln incinerators: This type of incinerator has wide application for different types of wastes, *i.e.* sewage sludge, municipal solid waste, hazardous waste, industrial waste, or for cleanup of contaminated soils (Dhote et al., 2020). It is also a two-stage incineration similar to starved air incinerator but the oxidative conditions with excess air of about 50-200% is operated at first stage rather than the semi-pyrolytic mode. The primary chamber of rotary kiln consist of an inclined cylinder lined with ceramic material, which moves on rollers at a revolution of two to six per hour based upon waste and rotary kiln type. The size of the kiln is around 1-6m in diameter and 4-20m in length. A series of rollers are used to rotate the kiln. The waste is loaded at the front end and ignited by a burner. The combusting wastes mixture is agitated in rotating kiln and it slowly tumbles down the chamber and drop down as ash. The mixing exposes the surface of the waste to heat from auxiliary burners. The kiln operates typically between 900 and 11000C, but high temperature of about 15000C is used in '*slagging type*' of rotary kiln that results in the formation of molten slag of ash and whole drum containing waste is incinerated. The molten slag retains the particulate matter, including heavy metals and forms a glass-like material which is less vulnerable to leaching of the dissolved metals when disposed into a quench pit. The gases from the primary rotary kiln shift to the secondary chamber, where excess air with auxiliary burners completely burn the combustible gases, vapors, soot, and tars. The rotary kiln provides a high temperature, and extended residence time which permits the complete combustion of mixed hazardous wastes.

Plasma gasifier technologies are considered to be well established, particularly for the vitrification of incinerator ash. Plasma is one of the four fundamental states of matter, *i.e.*, liquid, solid, and gas. The gaseous state of matter is heated or subjected to a strong electromagnetic field with a laser or microwave generator at temperatures above 5000°C to create plasma. This low density ionized vapours are used in plasma gasifiers to heat the feed stream. The main gases that are used as plasma gas are air, argon, nitrogen, carbon dioxide and steam. The liquid and solid feedstock materials are converted into vitrified slag and synthesis gas using extremely high temperatures (about 930–20,0000C).

Advantages and disadvantages of incineration

Incineration can significantly reduce the volume of bulkier wastes, thereby making their management simpler. All combustible carcinogens, mutagens, and teratogens, including pathological wastes, can be detoxified through incineration. There is no leachates generation, landfill gas emissions, and long-term odor as that associated with landfilling. Air emissions from incineration can be appropriately managed and controlled at a central location as opposed to widely dispersed landfill locations. Further energy recovery of the heat released can generate income and contribute electricity generation in the country's economy. However, there are some drawbacks of this technology like that of high initial capital investment, maintenance costs, compliance with rules and regulations, need of highly trained operators, qualified supervisors, and operating personnel. The emission of odours, particulate matter, and hazardous gaseous substances like CO, SO_2, NOx, heavy metals, ashes, *etc.* associated with incineration must be appropriately controlled to keep these undesirable air pollutants from being entered into the atmosphere. Many times, the siting of hazardous waste incinerators has become a highly controversial issue from citizen reluctance for establishing the incinerator near there settlements.

5.2.3.3 Some case studies related to Incineration

Ashden Awards for Sustainable Energy

The two Indian firms have won the prestigious Ashden Award for sustainable energy in 2011, which is considered as the "Green Oscar". The then chief of the Abellon Clean Energy, Gujarat, Mr. Pankaj Patel, received the award on behalf of the company for making biomass pellets from crop waste. These pellets fuel Gujarat's industries and provides farmers a market for their agriculture waste. Mr. Gyanesh Pandey, on behalf of Husk Power Systems (HPS), Bihar, has received this prestigious award for sustainable energy. The HPS has installed more than 65 power plants across the state operated through rice husk gasification, embedded local grids, and employed native people to help run the plants (The Economics Times, 2011). This has replaced diesel and kerosene as a power source. The cost of this energy, is only Rs 100 per connection to the people. Each power plant installed by HPS not only provides employment to the locals (about 300) but also serves about 400 households by replacing some 18,000 litres of diesel, and 42,000 litres of kerosene every year. The growth of husk power is very rapid, and it is estimated to proliferate over the next four years and envisions adding over 300 mini-grids in India and Tanzania, installing 15 MW of 100 percent renewable power plant assets that would remove 150,000 tons of carbon dioxide per year. These mini-grids will serve over 100,000 consumers (HPS, 2018).

Waste-to-green energy plant in Hadapsar (Pune Municipal Corpo-ra-tion)

According to reports of 2017, this plant now wears a deserted look as it has simply become a disposal site without production. The plant's backyard, which is near residential areas, is causing odour and water pollution problems (Chavan, 2017). The plant on a build-operate-transfer basis having capacity to process 700 tonnes dry waste operated by Mumbai-based Rochem Separations Systems (India) Pvt Ltd, was failed to run to its full capacity. This plant was planned to generate 10 MW electricity every day but, it has led to the stacking up of the rejected material and garbage being disposed in the open in the plant's backyard, near to residential areas. The Maharashtra Pollution Control Board (MPCB) gave the company the authorization in May 2011 for the generation of electricity through gasification/pyrolysis technology from waste. In 2010 the contract to generate electricity from waste was given to the company by the PMC but, reports published in foreign press about the gasification technology used raised doubts. The functioning of Ramtekdi plant was affected and it did not run to its full capacity and the civic body also doubted on electricity generation. However, the visit report (March 15, 2013) by MPCB officials mentioned the production of 1.5 MW electricity that was used to run the plant, which is somewhat contradictory.

The 1987 Timarpur Incinerator Experience

The Timarpur Refuse Incineration-cum-Power Generation plant was designed and built by the Volund Miljotecknik Ltd. Denmark. The investment cost of plant was Rs. 20 crores (US$ 4.4 million). The plant was setup by the Ministry of Non-Conventional Energy Sources (MNES) in 1987 and designed to burn 300 tons of municipal solid waste (MSW) per day to produce 3.75 MW of electricity. The plant works for 21 days of trial basis and shut down soon due to waste poor quality. The supplied wastes were having the calorific value of only 600-700 kcal/kg instead of the desired calorific value of 1462.5 kcal/kg. The plant was even tried to operate on diesel but failed. Following this, the Delhi High Court ordered an inquiry by the Comptroller Auditor General (CAG). The CAG in his submitted annual report in 1990, reported that the said plant remained inoperative since its installation. The ministry failed to utilize or dispose of the non-functional plant and incurred an expenditure of Rs 1.25 crore (US$ 278,000) on plant maintenance and insurance. The project was formally scrapped in July 1990.

5.2.3.4 Composting

Composting can be an excellent option to manage biodegradable or organic wastes; however, this method occupies space and time. This method is often promoted as a natural process of SWM. It is an aerobic process, where the end product is the partially decomposed organic fraction known as compost. One of the first organized composting operations to be reported upon in literature was carried out in India in the early 1930s, commonly known as the Indore process or Sir Albert Howard's Indore compost method.

As per fundamentals, aerobic microorganisms through a series of exothermic reactions extract energy from the waste organic matter that breaks down the complex material to simpler. The basic aerobic decay equation for composting can simply be depicted as:

$$[\text{complex organics}] + O \xrightarrow{\text{Micro.org}} CO_2 + HO + NO - + SO_2 - + \text{heat} + \text{other less complex organics}$$

During this decomposition, the temperature increases to about 700C in most well-operated composting operations. As the reaction develops, the early decomposers are mesophilic bacteria followed after about a week by thermophilic bacteria, actinomycetes, and thermophilic fungi. Above 70 ^{0}C, spore-forming bacteria predominate. As the decomposition slows, the temperature drops and mesophilic bacteria and fungi reappear. Protozoa, nematodes, millipedes, and worms are also present during later stages. The concentration of dead and living organisms in compost can be as high as 25%. The elevated temperature destroys most of the pathogenic bacteria, eggs (e.g., Ascaris) and cysts (*Entamoeba histolytica*). Size of waste, oxygen content, temperature, moisture, pH, C: N ratio are the main factors that influence the composting process and needed to be adjusted (Mandpe et al., 2021b).

The two main methods of composting can be classified as agitated and static. These methods are very much in use these days. The agitated method or windrow method, as name indicates, is periodically mixed for proper aeration to introduce air, to control temperature, and to obtain more uniform compost. Contrary to this is the static method (also known as static pile method) where the air is blown over the material to be composted and the waste remains static. The system in which composting operation is carried out in a reactor of some type is known as in-vessel composting systems (Negi et al., 2020; Mandpe et al., 2021a).

***Vermiculture*:** It is the rearing of earthworm for the decomposition of organic waste material. Earthworms, along with soil microorganisms, degrade organic waste materials and thus maintain the nutrient flux in the system. The waste of one organism is the energy source for another. The earthworm is physically an aerator, mixer, and crusher, chemically a degrader, and biologically a stimulator in the decomposer (Senapati and Dash, 1984). They play an important role in nature; hence they are called "friends of farmers," "ploughman of the earth," Cinderella of organic farming", "planets soil processors and builders," *etc.* They are a specific group of invertebrate animals belongs to the class *Oligochaeta* within the taxonomic phylum *Annelida.* They are cold-blooded animals and are hermaphrodite. Scientists so far discovered and named more than 4400 species of earthworms on this planet, and they can be divided into three basic categories of epigeic, endogeic, and anecic based on their ecological niche and habitat. Earthworms constitute 20% of invertebrate biomass in most the agricultural soils (Mandpe et al., 2020a). They degrade about 20% of NET primary production (NPP) and increase the composting activity by 20%. Earthworms contribute about 50-100kg of N_2/acre/year and possess enzymes urease, cellulase, which convert uric acid, fecal matter to nitrogen.

5.3 Scientific Challenges

Resource recovery from the waste is a very complex. This process involves many social, economic, and political implications associated with it that requires critical analysis and evaluation before a commitment is made. The process demands material quality, assured quantity of solid wastes, capital and operating expenses, the market value of retrieved materials, potential minimum reliable energy trade, persistent need for a sanitary landfill for excess and remaining unwanted materials disposal, and incinerator residue (Mandpe et al., 2021c). The main obstacles to resource recovery could be location, the uncertainty of supply, low value of the product, uncertain market, heterogeneity and putresibility of the waste, unverified technology, industrial and administrative constraints, and legal restrictions.

Techniques involved in size and volume reduction and resource recovery:

- Compaction: leads to volume reduction of waste mechanically
- Incineration: volume reduction
- Mechanical size reduction: cutting, grinding, and milling systems.
- Component segregation: hand-picking, air and magnetic separation, and screening.

5.3.1 Value Addition of Recyclables for Reuse

Value addition is a measurement of economic activity associated with recycling. It is measured by tracking the rise in prices paid for the recyclable materials as they progress through different stages of recycling (Weston, 1994). Value addition can increase either the product's price or cost. The value addition of recyclable items for reuse can be increased by removing the impurities and contaminants from the materials to be recycled. For example, buyers of waste lubricating oils can never be sure that they are free of specific contaminants like polychlorinated biphenyls, heavy metals, halogens, or water, which will make re-refining more difficult (Policy Brief, 2007). To increase the use and value of the product, producer can do some experiments. The mixing of flower petals and grass pieces into the recycled paper by Development Alternatives, an NGO in Gurugram, India, is one such example that increases its value and demand. Value additions in plastic wastes can be achieved by sorting different types of plastics and then sending them for further processing. Plastics from waste electrical and electronic equipment can be blended with virgin plastics to produce decorative glass panels, wood plastic panels, or acoustic panels. In Japan, recycled resin from PET bottle waste is used to manufacture civil engineering material and inner cotton for bedding. The value addition was made by colouring the recycled material at the resin stage, and high-quality carpets were made by using high-level sewing skills (Ministry of Environment, 2012). As a general rule, separation of color from the glass leads to efficient and sustainable use of glass cullet, and upgrading of the glass recycling rate with no mixed cullet surpluses that might have a negative impact on the market demand for recycled glass.

There must be recycling targets set by the concerned authorities for a growing range of materials. This approach can lead to "sustainable materials management," thereby helping to reduce pressures on the natural resource reservoirs, e.g., recycling of paper and cardboard packaging not only reduces the number of trees being cut down to generate pulp but also uses less energy than making paper using virgin materials. However, in the concept of value addition, the quality and cost of recycling paper should be equal to that of virgin paper so that it can be purchased and used by the user. Many times it can be hard for buyers of recyclable waste to find sellers of waste, and vice versa. Further, the market chain must be there so that the recycling industry can sell their products without many efforts and search costs from the buyer point of view can be reduced. Also, consumers may be hesitant to purchase products made from recycled materials due to lack of accurate information regarding their reliability and performance. This type of mistrust must be overcome by the concerned authorities either by issuing some standard symbols or certificates to the recycle product. Five issues have been identified as potentially significant barriers and failures in markets for recyclable materials as: information failures; technological externalities; search and transaction costs; consumer perceptions and risk aversion; and market power (Policy Brief, 2007). The initiation by MoEF&CC, Government of India through its draft policy on National Resource Efficiency NREP, 2019 is a good step in value addition through its guiding principles of waste minimization and material security. According to this policy, the manufacturers and service providers will be encouraged to use less virgin material and more recycled or renewable materials. Further, NREP targets 100% recycling and reuse rate of polyethylene terephthalate plastic by 2025 and 75% recycling and reuse rate of other plastic packaging materials by 2030.

Value addition will be only relevant and useful if problems of the recycling industry can be solved. Some common issues related to recycling are:

i) Generally, the present system of recycling in India and other developing countries is highly labor oriented with some organized recycling systems.

ii) The processing of segregated items is carried out by small-scale industries without compliance with environmental regulatory requirements.

iii) The poor state of the rag pickers' health and working environment.

iv) Unhygienic working conditions at the waste dealer (Kabariwala) complex and recycling factories.

5.3.2 Disposal of Process Waste and Residues

Process wastes are unavoidable leftovers residues or sinks for undesirable ingredients in the products. The quality of which mainly depends upon the process and the fuel used in operation. As per the Waste Incineration Directive 2000/76/EC European Parliament and Council (2000) residues means any solid or liquid substance (including fly ash and boiler dust, bottom ash, and slag, wastewater treatment plant sewage sludge, spent catalysts and spent activated carbon, solid reaction products from gas treatment,) produced by the incineration or co-incineration process, the exhaust gas or wastewater treatment or other processes within the incineration or co-incineration plant (Mandpe et al., 2019b). The residues need to be minimized in their volume and harmfulness and recycled accordingly. The residues which cannot be reduced, recycled or prevented can be disposed of as per regional and national rules. Due to high pollutant load boiler and filter ashes produced during incineration are characterized as hazardous waste and must not be mixed with the grate ashes. They are generally dumped of in exclusive and expensive underground disposal sites or a hazardous landfill. In Germany, they can be filled in old salt mines for backfilling of caverns (Vehlow et al., 2007). The two processes for their disposal are the Bamberg Model, where filter ashes are stabilized and neutralized in a landfill by combining with the wet scrubber sludge discharge (Rimann, 1990) and stabilisation with the Swiss filter ash cement after washing (Tobler,1989). In the same manner the leachate generated from landfills can be treated along with wastewater and landfill gas can be used to generate thermal or electrical energy or can be simply flared to convert it into carbon dioxide.

5.3.3 Eco-friendly Recovery Solutions

By doing waste segregation at source itself and by adopting 6 R's *i.e.* rethink, reduce, repair, reuse, recycle, and recover waste can be managed in a better way. This step is possible in Indian contest if waste generator, collector, management authorities and concern government effectively do their role. Non-functionality and carelessness will result in total failure of the project as observed in many cases in India. As discussed above in waste disposal methods metals, plastics, glass and paper can be separated either at site or at material recovery stations. After that compost, vermi-compost can be produced from biodegradable fraction of the municipal wastes and combustible part can be used to generate heat and produce electricity. There are many success cases in this regard but it's always suggested to adopt eco-friendly techniques. Instead of land filling or any disposal method, an eco-friendly solution of organic waste disposal is biogas production through anaerobic digestion. For instance a 500 kg/day food waste disposal was set up at Chittapur site in Gulbarga

District in Karnataka from which about 27 kg of LPG is obtained along with daily generation of 1000 litres of digested effluent. The effluent is excellent organic manure and can be used for plants growth and in fields (Babu and Mdhay Kumar, 2017) fertile clay as raw material (Giri Babu and Krishnaiah, 2018). Plastic waste can be used as a component of a composite construction material. It can be used as an aggregate of concrete and an inorganic filling material (Binici, 2013; Jassim, 2016). Further, plastic cement can be produced by mixing waste of polyethylene materials with portland cement (Jassim, 2016). The Sudokwon Landfill in Incheon city in the Republic of Korea is one of the largest, which showed a sustainable model for urban waste management. About 500 tons of food waste is converted into renewable energy daily in this biogas plant (Nam, 2017). This landfill gas power plant generates electricity enough for its 440,000 residents and about $50 million worth of electricity every year.

5.4 Organisation Challenges

5.4.1 Organization and Structuring Waste Management System

Capacity building is required in understanding the various provisions of different rules and regulations related to SWM. In addition to capacity building, proper implementation of existing rules and their compliance is required. Generation and proper utilization of funds are essential to make SWM a successful one. In the Indian context, there is a need for incorporating the private sector in waste management. There must be a focus on developing standardplans to solve common problems faced by city municipalities.

In a success case study in Saharanpur city of Uttar Pradesh, ITC Ltd. has started the *Sunhera Kal* Mission in October 2006, under its social development initiative in collaboration with Muskan Jyoti Samiti (a non-governmental organization) to manage 65 tonnes of generated daily waste. The project has developed a model that assists in recycling/reusing of recyclable and biodegradable waste that reduces the burden of landfilling. The collected waste is separated through secondary segregation at the waste management site into non-recyclable, recyclable and biodegradable waste. The organic compost is prepared from biodegradable waste by processing, the recyclable waste is sold to private vendors, and non-recyclable waste is transferred to landfills. About 15% (non-recyclable) goes to landfills and about 85% of the total waste is reused or recycled (Bhattacharjee and Sinha, 2015). The success lies in the capacity building workshop of the human resource enrolled in the project and awareness among citizens.

5.4.2 Training and Awareness

According to famous Chinese quote, "*the best asset for one year is to grow grains; the best investment for a decade is to plant trees; the best investment for a lifetime is to educate people.*" Thus the success of a project lies when concerned people are aware of it. The theories of reasoned action and planned behavior can be used as a framework in understanding, explaining, and predicting behavior for the management programs of solid waste (Desa et al., 2012). These theories are a helpful guide for designing involvement strategies to maintain or change a particular behaviour. The assumption is based on the hypothesis that attitudes of individual are directly related with their behavioural intentions. The theory of reasoned action views an individual's intention to execute or not to execute as an immediate determinant of the action (Ajzen and Fishbein,1980). The individual behaviour and attitude can be changed by creating awareness and providing necessary information to deal with the generated waste. By convincing people regarding their involvement and small actions can bring a remarkable change while dealing with environmental or social problems. Once people realised the importance that waste can be used as a resource, SWM will become smoother and effective.

5.4.2.1 Awareness and Education Programs in India Towards SWM

According to the *Father of Nation* Mahatma Gandhi, sanitation is more important than political independence (UNICEF, 2012). From the very beginning, parents told their children about self cleanliness, hand washing after going to the washroom and before meals. Once the children start going to school, the teachers take up this task and include cleanliness of teaching rooms and surroundings. Authors remember that there was an hour dedicated to school campus cleaning every Saturday for this purpose, and often, physical education teachers also dedicate their first 10 minutes of this class for this work. As grade increases, there is a sanitation campaign on Mahatma Gandhi's birthday on 2nd October every year. During college and university education, these activities were continued under National Social Service activities on Environmental Day celebrations on 5th June and Earth Day celebrations on 22nd April. Later on the directions of the Supreme Court, Environmental Education made a compulsory subject for all undergraduate courses. The course content of that subject also includes topic like waste management, environmental pollution, consumerism and waste products, human health and public awareness so that every student must understand about the importance of environment and contribute accordingly. The particular skills and knowledge acquired from environmental education would assist in modifying human attitude towards the environment (Ballantyne, 2006). These students having some knowledge and skills in environmental education are more encouraged to participate in environmental conservation activities, and plans (Tal, 2004) thus would develop new concepts for the solution of environmental problems. Several Indian universities and institutes have successfully implemented a green and sustainable campus by adopting cleanliness, paper recycling, composting and vermicomposting, rainwater harvesting, wastewater treatment, tree plantation, public transportation or bicycles use to reduce emissions, solar water heating systems and environmentally friendly buildings. Some examples are Atal Bihari Vajpayee Indian Institute of Information Technology & Management in Madhya Pradesh, Manipal Academy of Higher Education, Manipal, Mangalore University in Karnataka, Dibrugarh University in Assam, Institute for Financial Management and Research in Andhra Pradesh and Kalasalingam University in Tamil Nadu (Maniar, 2018).

At National level, to reward open-defecation free (ODF) and thoroughly sanitized Gram Panchayats (elected village-level-self-governing bodies), blocks, and districts the Government of India (GoI) under its Total Sanitation Campaign started "Nirmal Gram Puraskar" (NGP) or "Clean Village Award" an award-cum-fiscal incentive scheme on 2nd October 2003 (DDWS, 2011). The Gram Panchayats are promoted to provide full sanitation coverage to *anganwadis,* schools, and households. A Nirmal Gram (NG), thus, signifies an ODF village with all houses, *anganwadis,* and schools having sanitary toilets, along with awareness among people about the priority of keeping personal and community hygiene and clean environment. The rural people are motivated to build and use toilets for achieving the status of NG, and thereby abstaining from open defecation and attain proper sanitation at the household level. The Ministry of Drinking Water and Sanitation (MDWS), GoI started NGP awards in 2011 which was discontinued consequent upon the launch of 'Swachh Bharat Mission-Gramin' (SBMG) in October 2014 as per its notification in 2015. The SBM-G which focuses on making all villages in India ODF villages by 2019 has successfully achieved its goal. As discussed above, the Government of the Indian state of Haryana has also started Wealth from Waste project to manage solid waste in rural areas. Dr. Bhindeshwar Pathak, the founder of the non-governmental organisation '*Sulabh*' started community toilets for the poor people use during the 1970s in India. This significantly improved the living conditions for many people in India. The UN Human Settlements Programme (HABITAT) also encourages the Sulabh toilets use (Ramani, 2012). These schemes directly or indirectly promote awareness among citizens and motivate them towards cleanliness.

Conclusion

Expanding urban areas, non-segregation of generated wastes, complex nature of generated municipal wastes, improper implementation of existing rules, failure of existing waste disposal methods, lack of infrastructure, and funds are the significant challenges associated with municipal solid waste management in India. Further, our traditional beliefs and approaches like "*not in my backyard*" and "*away from the sight-away from the mind*," even towards self-generated wastes, are some hurdles in this regard. The revised Indian Solid Waste Management Rules [7] also stress on mandatory source segregation of waste, fixing responsibilities of the generators and integration of waste pickers and waste dealers in the formal system to make waste management more effective. As observed from different case studies, improper use and ineffectiveness of different available techniques of harnessing energy and wealth from the wastes causes more environmental costs than economic. Therefore, due attention must be paid to waste composition, geographical conditions, people's attitude, adoption of economically reasonable and environment-friendly disposal, and treatment options to make municipal solid waste management an effective one. In addition to these strategies, public-private-government sector co-ordination is a must to achieve success in municipal solid waste management properly.

ORCID

Hardeep Rai Sharma http://orcid.org/0000-0002-5130-5205

References

Ajzen, I., Fishbein, M. 1980. Understanding attitudes and predicting social behavior. New Jersey: Prentice-Hall.

Babu, G. R., Madhav Kumar, G. M. 2017. An eco friendly solution to the food waste disposal. AIP Conference Proceedings 1859, 020068.

Ballantyne, R., Connell, S., Fien, J. 2006. Students as catalysis of environmental change: a framework for researching intergenerational influence through environmental education. Environmental Education Research 12(3-4), 413-427.

Berhan city of Ethiopia. Journal of Environment and Earth Science. 4(5):98–103. UNICEF India website. 2012. Nirmal Gram: the path to total sanitation. Water, Environment and Sanitation. Available online at: http://www.unicef.org/india/wes_2924.htm (accessed 01 August 2012).

Bhattacharjee, S., Sinha, U. K. 2015. Making the best out of waste. Ideas for India available at: https://www.ideasforindia.in/topics/environment/making-the-best-out-of-waste.html (accessed 04 November 2019)

Bhide, A. D., Sundaresan, B. B. 1983. Solid waste management in developing countries. INSDOC State-of-the-Art Report Series–2. Indian National Scientific Documentation Centre, New Delhi.

Binici, H. 2013. Effect of aggregate type on mortars without cement. European Journal of Engineering and Technology, 1(1), 1-6.

Biswas, A.K., Kumar, S., Babu, S.S., Bhattacharyya, J.K., Chakrabarti, T., 2010. Studies on environmental quality in and around municipal solid waste dumpsite. Resour. Conserv. Recycl. 55, 129–134. https://doi.org/10.1016/j.resconrec.2010.08.003

Chavan, V. 2017. Waste-to-power plant turns into dumping yard. Pune Mirror, August 4, 2017 available at: https://punemirror.indiatimes.com/pune/civic/waste-to-power-plant-turns-into-dumping-yard/articleshow/59906789.cms (accessed 03 November 2019).

Department of Drinking Water and Sanitation (DDWS), 2011. Assessment study of impact and sustainability of Nirmal Gram Puraskar. Ministry of Rural Development, Government of India.

Desa, A., Kadir, N. B. A., Yusooff, F. 2012. Environmental awareness and education: A key approach to solid waste management (SWM) – A case study of a university in Malaysia. Chapter 5, Intech Open Science: 101-112. http://dx.doi.org/10.5772/48169

Dhar, H., Kumar, S., Kumar, R., 2017. A review on organic waste to energy systems in India. Bioresour. Technol. 245, 1229–1237. https://doi.org/10.1016/j.biortech.2017.08.159

Dhote, L., Mandpe, A., Paliya, S., Kumar, S., Pandey, R.A., Kumar, R., 2020. Characterization of distillery sludge for its thermal properties and ascertaining its utilization as a low-cost fuel. J. Clean. Prod. doi:10.1016/j.jclepro.2020.120872

ERM. 2004. Strategic planning guide for MSW management, published as CD, available at www.worldbank.org/urban/solid_wm/erm/start_up.pdf.

European Parliament and Council, 2000. Directive 2000/76/EC of the European Parliament and of the Council of 4 December 2000 on the incineration of waste. Official Journal of the European Communities, 28.12.2000, L332/91.

Giri Babu, S. V., Krishnaiah, S. 2018. Manufacturing of eco-friendly brick: A critical review. International Journal of Computational Engineering Research, 8(2), 2250–3005.

HPS, 2018. Husk Power Systems receives $20 million investment from Shell, Swedfund, ENGIE Rassembleurs d'Energies to scale renewable mini-grid business in Africa and Asia. Press Release, Jan, 15, 2018 available at: https://huskpowersystems.com/husk-power-systems-receives-20-million-investment-from-shell-swedfund-engie-rassembleurs-denergies/ (accessed 03 November 2019).

Jassim, A. K. 2016. Recycling of polyethylene waste to produce plastic cement. In: 14th Global Conference on Sustainable Manufacturing, 3-5 October 2016; Stellenbosch, South Africa.

Kumar, S., Bhattacharyya, J.K., Vaidya, A.N., Chakrabarti, T., Devotta, S., Akolkar, A.B., 2009. Assessment of the status of municipal solid waste management in metro cities, state capitals, class I cities, and class II towns in India: An insight. Waste Manag. 29, 883–895. https://doi.org/10.1016/j.wasman.2008.04.011

Kumar, S., Negi, S., Mandpe, A., Singh, R.V., Hussain, A., 2018. Rapid composting techniques in Indian context and utilization of black soldier fly for enhanced decomposition of biodegradable wastes - A comprehensive review. J. Environ. Manage. 227, 189–199. doi:10.1016/j.jenvman.2018.08.096

Kumar, S., Sharma, H. R. 2019. Wealth from waste an initiative for solid waste management in rural areas of Haryana: Issues, challenges and way forward. Indian Journal of Waste Management, 3(2): 61-67.

Kumar, S., Smith, S. R., Fowler, G., Velis, C., Kumar, S. J., Arya, S., Rena., Kumar, R., Cheeseman, C. 2017. Challenges and opportunities associated with waste management in India. Royal Society Open Science, 4: 160764, 11 pages.

Mandpe, A., Kumar, S., Lakshmikanthan, P., Hettiarachchi, H., 2019a. Mining for recovery as an option for dumpsite rehabilitation: Case study from Nagpur, India. J. Environ. Eng. Sci. 15, 52–60. doi:10.1680/jenes.19.00021

Mandpe, A., Kumari, S., Kumar, S., 2020a. Composting: A Sustainable Route for Processing of Biodegradable Waste in India. Org. Waste Compost. through Nexus Think. 39–60. doi:10.1007/978-3-030-36283-6_3

Mandpe, A., Paliya, S., Kumar, S., Kumar, R., 2019b. Fly ash as an additive for enhancing microbial and enzymatic activities in in-vessel composting of organic wastes. Bioresour. Technol. 293, 122047. doi:10.1016/j.biortech.2019.122047

Mandpe, A., Pratap, V., Lakshmikanthan P., 2021c. Landfill mining. In Current Developments in Biotechnology and Bioengineering (pp. 197-229). Elsevier. doi: 10.1016/B978-0-12-821009-3.00010-5

Mandpe, A., Tyagi, L., Paliya, S., Chaudhry, S., Motghare, A., Kumar, S., 2021a. Rapid-in-house Composting of Organic Solid Wastes with Fly Ash Supplementation: Performance Evaluation at Thermophilic Exposures. Bioresour. Technol. 337, 125386. doi:10.1016/j.biortech.2021.125386

Mandpe, A., Yadav, N., Paliya, S., Tyagi, L., Ram Yadav, B., Singh, L., Kumar, S., Kumar, R., 2021b. Exploring the synergic effect of fly ash and garbage enzymes on biotransformation of organic wastes in in-vessel composting system. Bioresour. Technol. 322, 124557. doi:10.1016/j.biortech.2020.124557

Maniar, P. 2018. Campuses across India take the green step forward. Hindustan Times, Jan 05, 2018, https://www.hindustantimes.com/education/campuses-across-india-take-the-green-step-forward/story-tItOpHDnIejuWlvrBsDYpK.html (accessed 03 November 2019).

Manyapu, V., Mandpe, A., Kumar, S., 2019. Corrigendum to "Synergistic effect of fly ash in in-vessel composting of biomass and kitchen waste" [Bioresour. Technol. 251 (2018) 114–120] (S0960852417321740)(10.1016/j.biortech.2017.12.039). Bioresour. Technol. 277, 225. doi:10.1016/j.biortech.2018.12.031

Ministry of the Environment, 2012. Solid Waste Management and Recycling Technology of Japan Toward a Sustainable Society. Minister's Secretariat, Waste Management and Recycling Department Policy Planning Division, Office of Sound Material-Cycle Society, 30 pages available at: https://www.env.go.jp/en/recycle/smcs/attach/swmrt.pdf (accessed 04 November 2019).

Nam, J. 2017. An eco-friendly approach to waste management. Development Asia, 5 pages. NREP, 2019. National Resource Efficiency Policy, (Draft). Charting a Resource Efficient Future for Sustainable Development. Ministry of Environment, Forest and Climate Change Government of India.

Negi, S., Mandpe, A., Hussain, A., Kumar, S., 2020. Collegial effect of maggots larvae and garbage enzyme in rapid composting of food waste with wheat straw or biomass waste. J. Clean. Prod. 258. doi:10.1016/j.jclepro.2020.120854

Paliya, S., Mandpe, A., Kumar, S., Kumar, M.S., 2019. Enhanced nodulation and higher germination using sludge ash as a carrier for biofertilizer production. J. Environ. Manage. 250, 109523. doi:10.1016/j.jenvman.2019.109523

Palnitkar, S. 2011. Millennium Development Goals and the Role of Indian Cities in Johnson, C (eds.) Indian Cities: Managing Urban Growth. A Metropolis Publication, pp.: 36-49.

Planning Commission, 2014. Report of the task force on waste to energy. Vol. I, Planning Commission, New Delhi.

Policy Brief. 2007. Improving Recycling Markets. Organisation for Economic Co-operation and Development (OECD), 8 pages available at www.oecd.org/publications/Policybriefs (accessed 01 August 2018).

Pratap, V., Bombaywala, S., Mandpe, A., Khan, S., U., 2021. Solid Waste Treatment: Technological Advancements and Challenges. In Soft Computing Techniques in Solid Waste and Wastewater Management (pp. 215-231). Elsevier. https://doi.org/10.1016/B978-0-12-824463-0.00014-8

Ramani, S.V., SadreGhazi, S. and Duysters, G. 2012. On the diffusion of toilets as bottom of the pyramid innovation: Lessons from sanitation entrepreneurs. Technological Forecasting and Social Change, 79: 676–687.

Reimann D.O. 1990. Reststoffe aus thermischen Abfallverwertungsanlagen, Beihefte zu Müll und Abfall, 22,12.

Senapati, B.K., Dash M. C. 1984. Functional role of earthworms in decomposer subsystem. Tropical Ecology, 25(2): 54-73.

Sharma, H. R., Destaw, B., Negash, T., Negussie, L., Endris, Y., Meserte, G., Fentaw B., Ahmed Ibrahime, A. 2013. Municipal solid waste management in Dessie City, Ethiopia. Management of Environmental Quality. 24(2): 154-164.

Sharma, H.R., Abebe, T., Admassu, M., Teshaye, T., Aseffa, T., Emana, M. 2011. Municipal solid waste management and community awareness and involvement in management and practice: an overview and a case study from Gondar town of Ethiopia. International Journal of Environment and Waste Management. 7(3/4): 294-304.

Solid Waste Management (SWM) Rules. 2016. The Gazette of India. Part II-Section 3-Sub-section (ii). Ministry of Environment, Forest and Climate Change Notification, New Delhi, No. 861, the 8th April, 2016.

Tal, R.T. 2004. Community-based environmental education – a case study of teacher-parent collaboration. Environmental Education Research, 10(4): 523–543.

Tchobanoglous, G., Theisen, H., Vigil, S. 1993. Integrated Solid Waste Management: Engineering Principles and Management Issues, McGraw-Hill, New York.

The Economics Times. 2011. Two Indian firms win Ashden Awards for Sustainable Energy. July 17, 2011 available at: https://economictimes.indiatimes.com/industry/energy/power/two-indian-firms-win-ashden-awards-for-sustainable-energy/articleshow/8887682.cms (accessed 03 November 2019)

Tobler H.P. 1989. Konzepte zur Reststoffentsorgung in der Schweiz, VDI Berichte 753, 9. Tyagi, V., Fantaw, S., Sharma, H. R. (2014). Municipal solid waste management in Debre

Vehlow, J., Bergfeldt, B., Wilén, C., Ranta, J., Schwaiger, H., Visser, H.J.M., Gu, S., Gyftopoulou, E. , Brammer, J. 2007. Management of solid residues in waste-to-energy and biomass systems. Forschungszentrum Karlsruhe in der Helmholtz-Gemeinschaft. Wissenschaftliche Berichte FZKA 7347, pp: 1-65.

Vikas. 2017. Solid waste management in Indian cities: Issues and challenges. International Journal of Advanced Educational Research. 2(5): 241-248.

Wainaina, S., Awasthi, M.K., Sarsaiya, S., Chen, H., Singh, E., Kumar, A., Ravindran, B., Awasthi, S.K., Liu, T., Duan, Y., Kumar, S., Zhang, Z., Taherzadeh, M.J., 2020. Resource recovery and circular economy from organic solid waste using aerobic and anaerobic digestion technologies. Bioresour. Technol. 301, 122778. https://doi.org/10.1016/j.biortech.2020.122778

Waste Atlas. 2017. Waste Atlas. Available online at: www.atlas.d-waste.com/ (accessed 23 May 2017).

Weston, R. F. 1994. Value added to recyclable materials in the Northeast. Report prepared for the Northeast Recycling Council, Council of State Governments, Brattleboro, Vermont.

World Bank. 2012. What a Waste - a Global Review of Solid Waste Management. Urban Development Series, March 2012, No. 15. Washington, DC, 20433, USA. Available at: worldbank.org/INTURBANDEVELOPMENT/Resources/336387e1334852610766/What_a_Waste2012_Final.pdf. (accessed 23 May 2017).

6

Economic Assessment of Waste Management in India

Alpana Gupta[a] and Athar Hussain[b]

[a]Alpana Gupta, Innovator, Indian Institute of Technology, Delhi
[b]Dr. Athar Hussain,, Professor and Head, Civil Engineering Department Room No. 104, NSUT West Campus, Formerly CBPGEC, Jaffarpur New Delhi-110073

ABSTRACT

Municipal solid waste management (MSWM) has become acute in India, due to the rapid urbanization and unregulated population growth rate. MSWM is provided low priority, although an important service. This service has not been satisfactorily served by lack of economic resources, institutional weaknesses, incorrect technological selection, and government apathy towards MSW. The present methods of uncontrolled waste dumping in cities have developed a severe issue for the environment and public health. Study will includes a publication to analyze and evaluate the current situation of MSWM in India, combining environment policy with the economy, opportunities, and limitations of linking economic data with a life cycle analysis. An approach to study the financial and environment friendly MSWM model to meet the future challenge is presented.

Keywords: Environmental Economics, Municipal Waste Management, Multi-criteria Evaluation Environmental Impact, Life Cycle Assessment, 3R Principle and Circular Economy.

6.1 Introduction

Urbanization is now a global phenomenon, but in developing countries the implications are more pronounced. Natural growth of population, reclassifications of habitation and migration trends are important in urban population in India. The population of India, which accounts for 27 per cent of the total population, was 285 million according to 2001 census. International experience shows that the pace of urban development accelerates when a country's urban population crosses close to 25 percent of the total population (like in India).(Kumar and Chakrabarti, 2010). Due to rapid urbanization and uncontrolled growth rate of population, SWM has become acute in India. SWM is established by municipal bodies in India. It is a vital service, however, but it does not achieve an appropriate priority and results are weak According to the Municipal solid waste management and processing rules -2000, the management of solid waste is mandatory for urban local authorities. However, the actual management of solid waste is given top importance and responsibilities are either not carried out or unwell achieved, which means the metropolitan faces many environmental and sanitary problems. The Indian approach to solid waste management is extremely inefficient, using an old, outdated system for the processing, treatment and disposal of storage collection as shown in figure 1. No formal organized biodegradable and non-biodegradable solid waste segregation system exists.

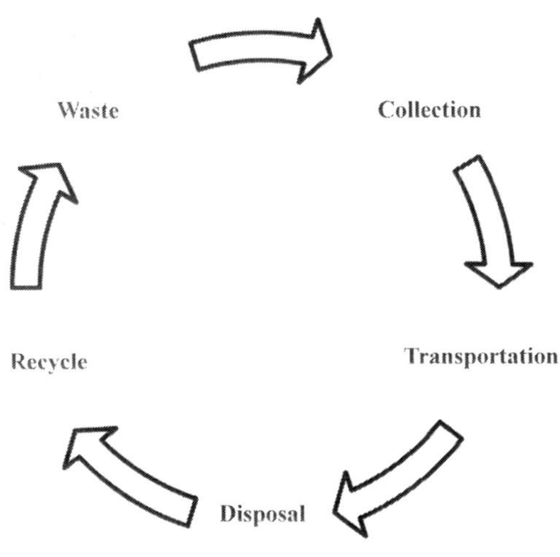

Fig. 1: Waste Management Process

Solid Waste generation

i. Residential solids, Institutional / Public areas

ii. Vegetable solid waste (retail $wholesale)

iii. Hotel and restaurant solid wastes;

iv. Residential solids:

v. Hospital biomedical waste and dispensaries:

vi. Industrial waste

Waste generation in Indian Urban Cities is indicated in Table 1.

Table 1: Amount of waste generation in Indian Urban Cities(Kumar and Pandit, 2013)

City	Generation of urban waste rates(Kg/Capita/Day)
Banglore .	53
Hyderabad	35
Hardwar .	40
Delhi	.60
Calcutta .	51
Sonepat .	343
Meerut.	45

6.1.1 Waste Treatment and Disposal Methods

Waste treatment and disposal methods are selected and used based on the form, composition, and quantity of waste materials.

i. Thermal Treatment

Thermal waste management refers to heat-handling systems of waste materials. ***Incineration*** is one of the most common waste treatment methods. It involves combustion in the presence of waste oxygen. The thermal treatment process is commonly used for energy and heat recovery. ***Gasification and pyrolysis*** are both comparable options of decomposition of biological waste, both by low oxygen and very high temperature exposure waste. Open burning is an old, environmentally hazardous treatment of thermal waste. ***Open Burning*** is a conventional thermal waste process that is unsafe for the environment. In the incinerators used in this process, there are no pollution control devices.

ii. Dumps and Landfills

The most common waste disposal solution is provided by sanitary deposits. These sites wish to eliminate or reduce the risk of waste disposal to the environment or public health (Mandpe *et al*., 2019a). The dumps are more and fewer the same as sanitation dumps regulated. Such dumps fulfill many of the criteria for sanitary waste, but there may be one or two lacking.

iii. Biological Waste Treatment

The aerobic composition of organic wastes is another method of waste disposal or treatment most commonly employed by small invertebrate and microorganism behavior. Statically stack composting, winding composting and in-vessel composting are the most common composting techniques. Anaerobic Digestion also uses organic decomposition processes. However, there exists oxygen-free world for anaerobic digestion for decomposing surplus materials in which composting needs air to enable the growth of microbes (Kumar *et al.*, 2018; Manyapu *et al.*, 2019)).

6.1.2 Issues and Challenges

Scientific and systematic waste storage is not practically possible in most cities of India. The waste is usually thrown into nearby vacancies, public vacancies, drains, streets, etc. The environment is hideous and hygienic because of waste thrown on the street, so that municipal workers are no longer allowed to clean the city even when they are regularly cleansed for 2 or 3 hours.

i. In general, people in residential, institutional, and commercial areas do not provide adequate bins at resources.

ii. In general, people do not take waste at the points indicated where it is carried to neighboring highways, railway tracks, open plots etc. There is no different system to keep biodegradable or non-biological waste.

iii. In most cities there is no waste processing. Very few cities are organized and administrative–for treating the waste in a very narrow measure, such as composting. Most of the waste is dumped in an open dump by the agency concerned without going to the site or waste details. No rules or norms on disposal are observed and the sites are not managed scientifically.

iv. Land use is amongst the most scientific and unhygienic practices in most Indian cities with severe environmental consequences. The waste is transferred and dumped at the site.

The most popular disposal method is the open dumping of waste at sites in Bangalore. Bangalore has 14 of these plants at which solid waste is dumped without processing or treatment at open heaps. There are two composting stations that process 200 SW metric tons daily and 300 SW metric tons daily, but the capacity of both plants is much lower and only one–a third of capacity can be processed. Solid waste is disposed of by landfilling and open dumping in Calcutta–in addition to small private landfills, more than 40 such disposal grounds exist. These sites receive solid waste from 500 to 10 tons per day. In Delhi, approximately 5,339 bins have been placed in two colors, in Blue & Red shades of non-biological and biologically degradable waste, however, subsequently human don't take waste seriously and waste is thrown out of the basin, there is a growing problem of solid waste every day. Private companies in Faridabad (Haryana) and Chennai are granted contracts by their respective municipalities to generate marketable solid waste compost there. Private companies are given incentives including lending land, insuring a fixed amount of waste every day, subsidies for purchasing equipment or tax vacation for definite phases on the revenue from the trade of the goods from the site.

6.2 Origination of Economic Assessment of Waste Management

The municipal waste management affects the population in several ways: in economic terms, by collections, in environmental terms and by emissions and indirect systematic effects but also by the physical connection with waste management through the design and psychology of the waste management system localization (i.e., NIMBY). All these aspects have to be taken into account by the decision-making apparatus. (Reich, 2005) LCA offers a useful tool to analyze municipal waste management in terms of its environment (i.e., system perspective, completeness, accepted framework, etc.).

Waste management is a directly related topic to conserving the environment and conserving resources and administrators and planners can become increasingly concerned with a healthy understanding of an integrated approach for waste management. (Costi *et al.*, 2004). Therefore, a well specified DSS is typically used to select the best options that can include decision support models.

6.3 Relationship Between Environmental Policies and Economics

The theory of cost and benefit spill (externalities) was first formulated by Henri Sidgwick (1837-1900). The definition of externalities was formalized by Pigou (1877-1959). He was a British welfare economist was concerned about how to use economic theory to promote social well-being. "." Externalities can be

(i) Positive or spill-over effect

(ii) Negative or spill over benefit

i. Positive Externality

Pigou highlighted the concept of a forest-planting exemplary spill over benefits. Around the property owners benefited from reforestation by natural seeding of their vacant land, yet no payment for the gain has been paid. In turn, the population had to do less planting. The positive effect of an event on an unknown third party is therefore a positive externality. For example, airport building and service. Due to increased accessibility, local companies would benefit (Verma, 2016).

ii. Negative Externality

Detrimental externality is an event that has a negative impact on third parties without a relation. It is a discharge of an economic transaction that negatively affects a party which does not participate directly in the transaction. For example, pollution is a negative externality. There are two types of negative externalities which are; *negative external consumption-* If an individual's intake decreases the health of others without compensation. The consumption of cigarettes reduces the pleasures of non-smokers who consume their food in the same restaurant in a restaurant which can allow smoking, for instance. *Negative income* — when the development of a business decreases the wellbeing of others not being paid by a company. To order to reduce emissions from nearby residents, the development of smoke from factories could, for example, generate clean-up costs.

6.3.1 Methods for Correction

***Tort scheme**:* This requires the government that determines the rule of tort and the rules. Tort laws are for any misdeed that leads to loss or harm which results in legal liability.

***Regulations of the Environmental Agency**:* When tax is cheaper than an alternative production method, the contamination issue shall not be resolved. Agencies such as CPCB must therefore follow administrative approaches

Suppressing Costs: Suppressing a good's external costs is a good way of responding to the externalities problem. The real and lower social cost and price is accomplished by internalization of costs.

6.3.2 Principles of Environment Law

The Doctrine of Precautionary and Polluting State Policy was made indispensable for sustainable development in 1996 after the Vellore Citizen's Welfare Forum vs. Union of India. The' polluter pays' rule is the common practice in which polluters should bear the cost of handling emissions to prevent harm to human health or the environment as shown in figure 2.

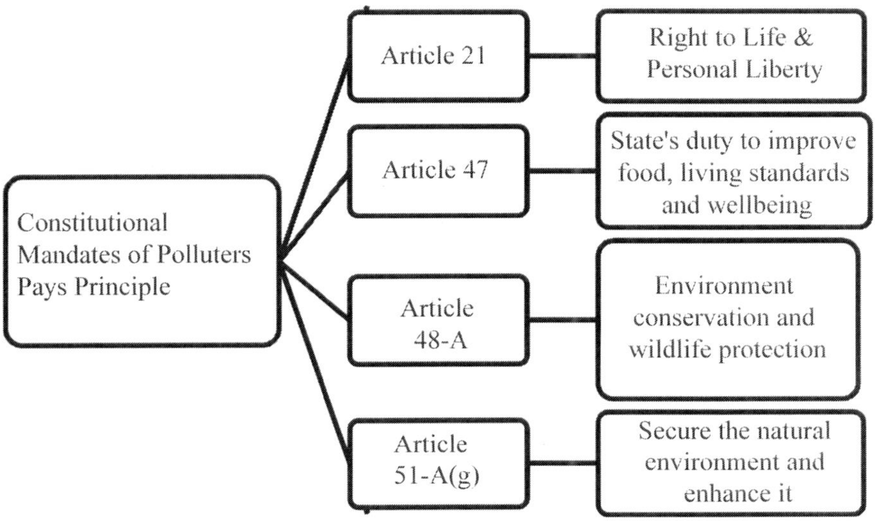

Fig. 2: Polluter Pay Principle

The ' polluter pays ' theory is a commonly accepted method of the expense of handling the emissions to mitigate harm to human health or the environment for those who create the pollution. There was a mistake. The majority of pollution control affecting land, water and air is based on this principle.

6.4 Environmental Principles & Economic Mechanisms

When the externality is negative, there is no damage to the first party for their social impact, while the second party is not unreasonably affected. If the polluter does not cover various costs the' polluter pays' principle is thus compromised. The principle does not imply' fault;' it encourages a' curative' approach to the restoration of environmental impairment. The harm that is critical to sustainable development can thus be remedied

i. Taxation Mechanisms

Emissions tax is part of the economic process aimed at reducing and adversely affecting emissions. There was a mistake. Standard means the amount of contaminants allowed during the taxable duration to be released to the atmosphere.

ii. Pigovian Tax

A Pigovian tax is a tax on all operations on the market that generate negative externalities .Pigovian taxes were named Arthur Pigou of England ,a theorist who also developed the concept of external economies. Negative externality occurs if the social cost is more than the private value of the margin (MSC >

MPC). The tax was intended to offset negative externalities and rectify markets so that production is effective.. *(pigouvian-taxes)*as shown in figure 3.

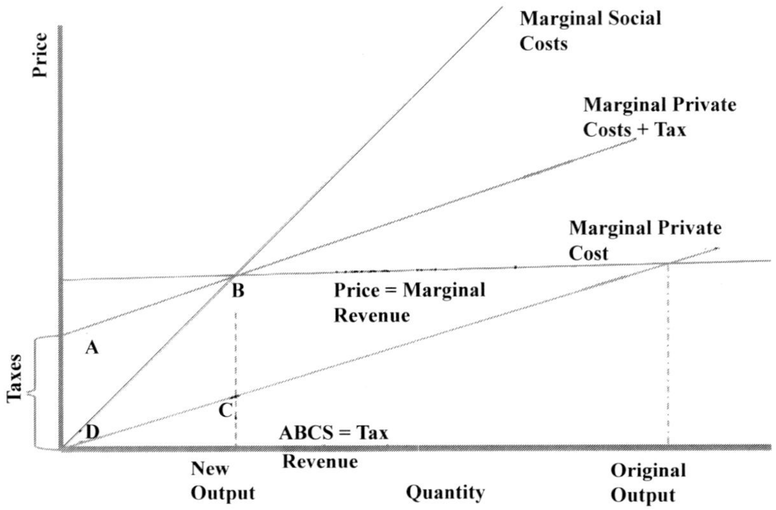

Fig. 3: Working of a Pigovian tax

A tax shifts the marginal private cost curve up by the amount of the tax.

6.4.1 Eco Taxes

An eco-tax is a tax levied on a product or service to suppress some or all of the peripheral costs of the operation performed or a tax on the use of protection against the environment. Taxes unswervingly discourse market let-downs by "charging rate on" ecological costs: High enticements for revolution: taxes raise polluter costs, give companies opportunities to create and implement existing technologies. Environmental tax classifications in four basic categories: energy, transport, emissions and resource tax as shown in figure 4.

Emission and Effluent Taxes	**Product Taxes**	**Natural Resource Tax**
No Taxes: Classified under this category in India	Clean Energy Cess Vehicle Entry Tax Vehicles Tax Ecological Fund and Environment Cess	Forest Development Tax

Fig. 4: Classification of Eco-Taxes

i. Raising revenue

In order to increase income primarily to fund public expenditures, governments often impose specific carbon and other taxes on economic bases. Most taxes related to the environment (e.g. for motor fuel and motor vehicles) are prime candidates for taxation as taxation appears to have a limited effect on short-term demand. The revenue-raising target could lead to environmental tax rates higher than the estimated value of social externalities.

ii. Double-dividend hypothesis

In the environmental sense, Tullock (1967) put it forward in a paper called, Excess Profit, that taxes on polluting activities that bring two kinds of benefits as shown in figure 5.

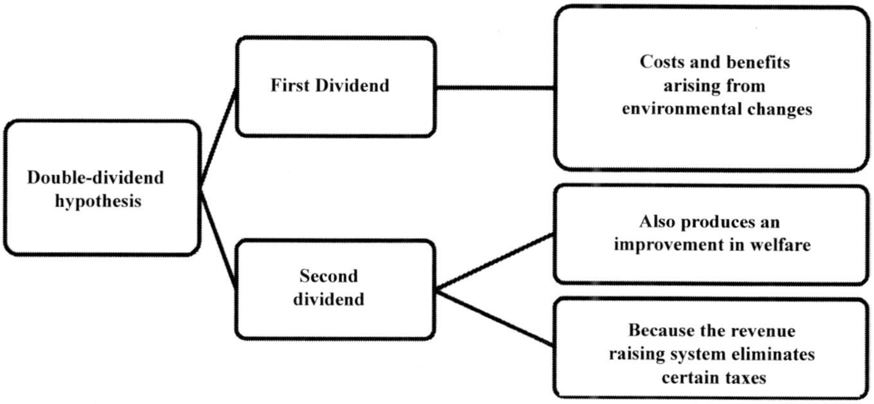

Fig. 5: Double-dividend hypothesis

iii. Shortcomings of eco taxes in India

In the case of India, most of these taxes didn't have a progressive rate structure even though some of them had an environmentally relevant tax base as shown in figure 6.

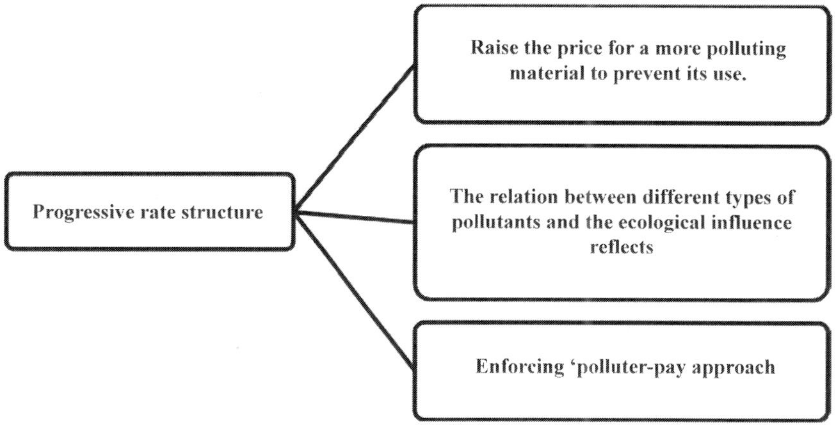

Fig. 6: Shortcomings of eco taxes

iv. Drawbacks of Polluter pay principle

The polluter pays rule does have a positive impact on emissions control. It is real. Nonetheless, the issue of "who's a polluter" still persists. In legal terms the definition is as broad as it becomes unsupportive in many situations. The size of the industry must also be considered so that the penalty can be graded accordingly. There is no gradation system prescribed on the amount of pollution or degradation caused by an industry. There cannot be any additional charge for power or waste disposal for a significant number of deprived families, casual segment trades and sustenance farming. (Johnson, 2017)the world's biggest waste-to-energy (WTETraders in rising republics can not necessarily transfer the price internalisation burden to foreign customers due to flexible demand.

6.5 Models for Waste Management

Different waste management classifications have been made and several reviews and analyzes have been classified as follows:

a. models based on cost-benefit analysis,

b. models based on life cycle analysis,

c. models based on multi-criteria decision analysis

i. Cost Benefit Analysis

Cost-based benefit- analysis models (CBA) are seen as optimization replicas that use a range of scenarios for choosing optimistic and deleterious effects as monetary effect (the environmental impact not monetized should be monetized)(Behzadian *et al.*, 2010) .According to Hansjurgens (2004) CBA contains eight stages as shown in figure 7.

i. The definition of the project proposed (including resources for start-up and progress)

ii. Addresses the identified project impacts since their minimization involves resources that are necessary for environmental and health protection;

iii. Identifies the relevance of the economic point of view identified impact;

iv. Assess impact-related effects;

v. Represents the corresponding monetary assessment of the impact;

vi. Includes cost and benefit expression in the current-current values and future cost discount;

Defination of the Project

↓

Identifying project Impacts

↓

Economic relevant Impact

↓

Phsical valutaio of relevant effects

↓

Monetary Valuation

↓

Cost and Benefits flow

↓

Net Present value test

↓

Senstivity test

Fig. 7: Stages of CBA

ii. Multi-criteria decision support derived prototypes

MCA is involved in the compromise methods group measured in simulating waste controlling. Multi-criteria decision analysis (MCA) It is used to assess environmental problems through the development of a full range of alternatives based on various criteria. (Morrissey and Browne, 2004) The MCA model type of criterion selected depends on the project goals.(Kapepula *et al.*, 2007).

- *MCA-based model models*

The type of criteria selected depends on the objectives of the model. Several types of MCA model have been developed, for example, to help detect the best compromise alternative. In conjugation with the ARGOS technology, the PROMETHEE model has been used to compare certain regions of certain cities with three main aspects in the fields of solid waste management (production, waste collection and handling) which are focused on nine parameters: Surplus per household volume, estimated host number, duration of two collections, use of the collective resources, household resistance to taxes on the management of solid waste, frequencies of collection, distances by household, cultural resources for eliminating solid household waste, perceived household nuisance.

• *Life Cycle Assessment framework*

A précis mode of assessing the ecological effects of a produce, procedure or structure, from acquisition of the raw materials through production, distribution and utilization, possible reuse / recycling and final disposal over a whole life cycle is the Evaluation of the life cycle (LCA) as shown in figure 8.

• *Life Cycle Assessment framework*

The detailed LCA follows the boots as specified by standards of ISO (14040:1997): (Garbie, 2015).Target Description and range, target description, scope and device limitations. Inventory review with a comprehensive list of all industrial and ecological inputs (recourse and flow of fuel and resources) and outputs (emissions and waste); Interpretation–Analysis of stock evaluated.

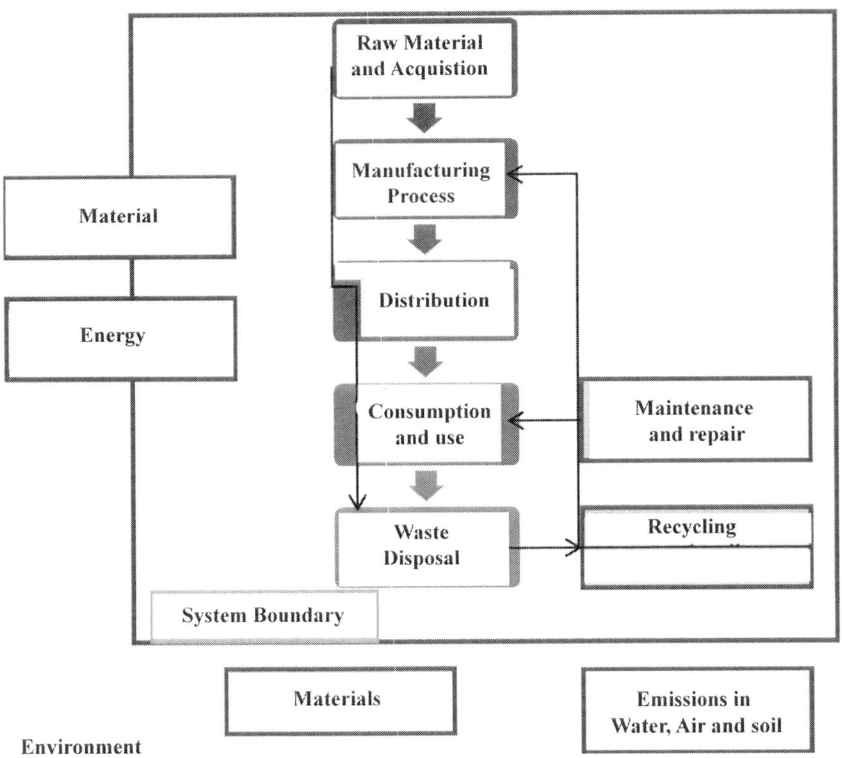

Fig. 8: *Life Cycle Assessment framework*

For the assessment of effect categories in the LCA evaluation, the fallouts of a life cycle inventory are used which include four steps:

- Collection of impact classes and sorting: relevant categories of study are defined
- Characterization: stage for simulating for each type of emission in harmony through environment discipline.
- Standardization: impact ratings and resource consumption are standardized using a mutual locus to compare influence classes.
- Evaluation: evaluation allocation of different environmental impact groups reflects their relative importance in the analysis.(Banar, Cokaygil and Ozkan, 2009)

The categories for impact can be evaluated in the life cycle assessment phase, including global warming potential, depletion of abiotic resources, climate change, human toxicity, acidification, and eutrophication.(Hansen *et al.*, 2006) Characterization factors for impact assessment of emissions are presented in Table 2.

Table 2: Characterization factors for impact assessment of emissions (Bovea and Powell, 2016)

Performance indices	Impact class summary	Contributing to the class of effects	Formula	
Global Warming	Is the influence of the lower atmosphere rising temperature	CO_2, CH_4,N_2O, CFC, HFCs	$GWP = \sum_{i=1}^{n} GWP_i \times m_i$	mi – the material mass i released in kg GWPi-the substance's global warming potential.
Ozone Layer depletion	The stratospheric intake of anthropogenic halogenated compounds enhances the decomposition of ozone.	CFCs, HCFCs	$ODP = \sum_{i=1}^{n} ODP_i \times m_i$	MI- is the substance emissioni;i; ODPi- is the material equivalence factor i.
Photochemical ozone formation	Once nitrogen oxides are detected, ozone is produced in the troposphere under the influence of sunlight.	NOx, VOCs including CH_4,CO	$POF = \sum_{i=1}^{n} POCP_i \times m_i$	POCPi – the capacity for photochemical ozone formation of the substance i; mi – the substance i emissions.
Acidification	There are several pollutants that are either acidic or converted into acid by air processes.	SO_2, SO_3	$AP = \sum_{i=1}^{n} AP_i \times m_i$	APi- material acidification potential,i I mi-substance emission i.
Eutrophication	Include all potential environmental effects of macronutrient concentrations.	NO_2, NH_3,	$EP = \sum_{i=1}^{n} EP_i \times m_i$	EPI – The substance's eutrophic potential i mi – substance's emission i.
Human toxicity	Negative health effects of naturally emitted toxic substances.	VOC , particles, heavy metals,	$EP_{hum} = \sum_{i=1}^{n} EQF_{i,m} \times m_i$	EPhum, m-the possible impact (humantoxicology) of micropathy, EQFi, m — the substance i caused by exposure through micropathy (water, soil, air),mi — substance emission i.
Abiotic depletion	Natural resources such as iron ore and crude oil are abiotic resources.		$ADP = \sum_{i=1}^{n} ADP_i \times m_i$	ADPi abiotic depletion Resource potential i, mi – resource amount i extracted

6.6 Tools and Techniques for Effective Waste Controlling

Waste reduction is a collection of processes and practices to reduce the quantity of waste produced. Waste minimization requires a redesign of goods and systems and/or improvement in consumer use and output patterns. The most environmentally sound, economically efficient, and economically viable way of waste management often is to reduce or eliminate the generation of harmful and persistent waste. waste minimization promotes efforts to promote a more sustainable society.

i. Economic Valuation

Waste management solutions can be analyzed in terms of privately cost. The methods for waste clearance are calculated on a gross cost of each ton of waste and recycling schemes can be analyzed on a financial basis. In its management strategy, high priority would be given to systems with the highest net efficiency and/or lower cost disposal options. In the past, 19 environmental and economic precautions have been barely noted in practice. Verdicts on managing waste should no longer be justified merely by comparison to economic charges and profits. A number of social and environmental impacts should be taken into account in the evaluation process. This can be accomplished through a thorough economic assessment of waste management solutions that includes negative externalities cost and societal merits. This is usually done by way of a social cost-benefit analysis.

ii. Social cost-benefit analysis

A social cost / benefit study includes the drawbacks, advantages and environmental benefits of the disposal of waste process. In line with the polluter pay principle, all the social costs of production and consumption must be expressed in producers 'production costs and market prices for the consumer. When these service societies set their values, greed's came to know the actual importance of nature as nature is not a free good(Samiha, 2013). Following are the variables required for smooth assessment financial function of environmental services:(1), market creation, by using social surveys, to determine the willingness to pay by consumers for natural goods i.e. contingent evaluation (2) Identifying consumer replacement markets that serve as market regulator for lack of environmental goods or services (e.g. market-based hedonic pricing); (3) Establishing dose-response relationship i.e. the link between the pollution level and damage or losses caused; finally

(4) An implicit social interest or quality can be defined with a legal and/or regulatory approach. It was justified to encourage awareness, to justify a decision, to assess the rules, so that macroeconomic objectives were taken into account and (rarely) compensation was determined.

6.7 Concept of 3Rs

There is often a debate on the origin of the sentence "reduce, reuse, recycle" However, it can be traced back to the underlying environmental awareness movement in the 1970's. During the Vietnam War, Americans demanded that attention should be devoted to air pollution, waste and water quality.

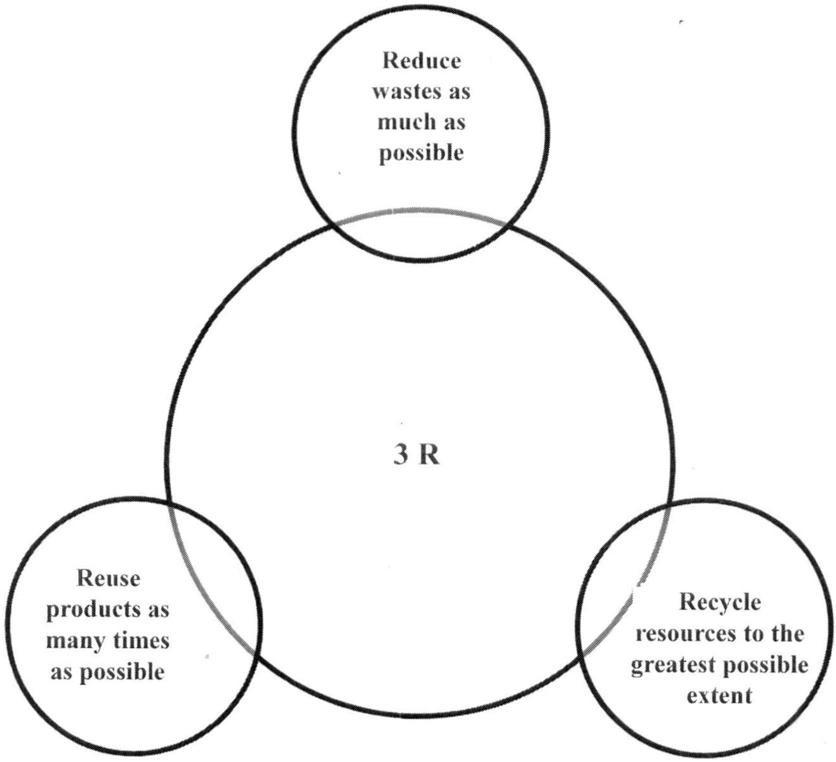

Fig. 9: Concept of 3Rs

UNIDO launched an initiative for green industry that sets forth a world where industrial industries reduce waste in each form, use renewable resources as input materials, and take all possible precautions to prevent harm to workers, the climate, or the environment.–ESSENCE OF THE 3Rs APPROACH as shown in figure 9.

i. Advantages

- Pollution Minimizes-All forms of pollution emanate from waste in the modern world.

- Protects the environment-Play a big role in protecting Mother Nature in the most balance. Reduction & recycling of this waste goes a long way towards significantly reducing pollution levels. While many trees are cut daily, recycled paper produced from specific trees is used continuously to reduce deforestation.
- Cutting waste at sites–it is beneficial because it contributes to minimizing soil and water pollution, because sites contribute greatly to the degradation of the environment.
- Low carbon footprint, reduce energy consumption, and create jobs.

i. Disadvantages

- High upfront cost of capital–recycling is not always economic. The construction of a new waste recycling unit requires a lot of money. The costs are to purchase various types of utility vehicles, upgrade the recycling facility, waste, chemical waste disposal, and educate the local residents through useful programmers and workshops.
- Recycled waste products may not be long-lasting. –The quality of recycled waste products may not match the quality of the waste as the products are from overused and fragile waste materials.

6.7 3Rs Technologies and Resource Efficiency

The provision of reasonably priced products and services that meet consumer needs and provide quality of lives while reducing slowly ecological impacts and resource levels slowly over the life cycle at least to a rate that is compatible with the Earth's projected carriage potential. Following are the constituents needed for achieving Eco-Efficiency:

i. By gradually decreasing material concentration

ii. Decreasing rate of energy being utilized

iii. Reducing the concentration of emission into the atmosphere.

iv. Enabling recycling environment

v. Potential usage of renewable resources

vi. Increased rate of product life

vii. Serviceability should be increased

Only by dematerializing economic activities can sustainable development be approached. Productivity of resources focuses on new technologies to reduce inputs of materials and generate the same or even better output services.

Identifying areas are responsible for the extraction and consumption of material, energy, and land.

6.7.1 China – 3Rs- Circular Economy

The core principle of China's circulatory economy is ' 3R ': reducing resource consumption and waste / pollutant generation during manufacturing, circulation and consumption. Recycling of the waste as product directly or post repairing, renovation, or re-use as part of other product, wholly or in part.

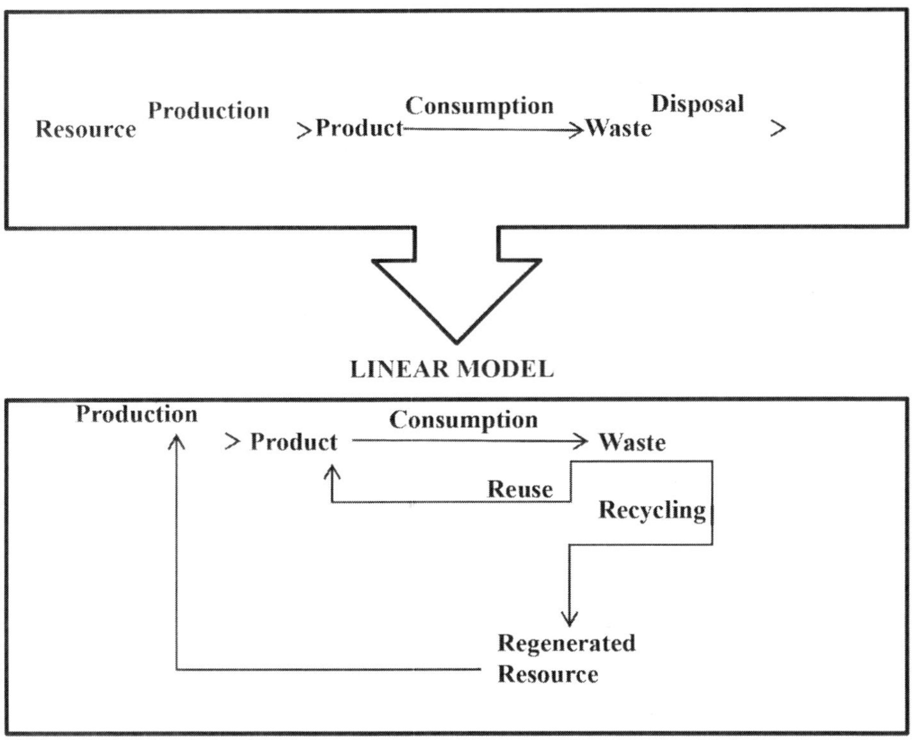

Fig. 10: A 3R Implementation Framework to Enable Circular Consumption in Community

The development of a mechanism for implementing circular theory based on the reduction-reuse-recycle (3R) concept, the promotion of circular consumption in the society and acceleration of pro-3R modifications, is an indispensable component in the circular economic system to promote economic growth and to reduce the degradation and depletion of environmental assets.3R-Shop offers low-carbon marketing, shipping and trade second-hand products, and storage and selling of domestic waste .For example; 20 community shops has adopted the sustainable way of disposing and reusing waste in Wuhan, China and 3R implementation framework is shown in figure 10 and figure 11.

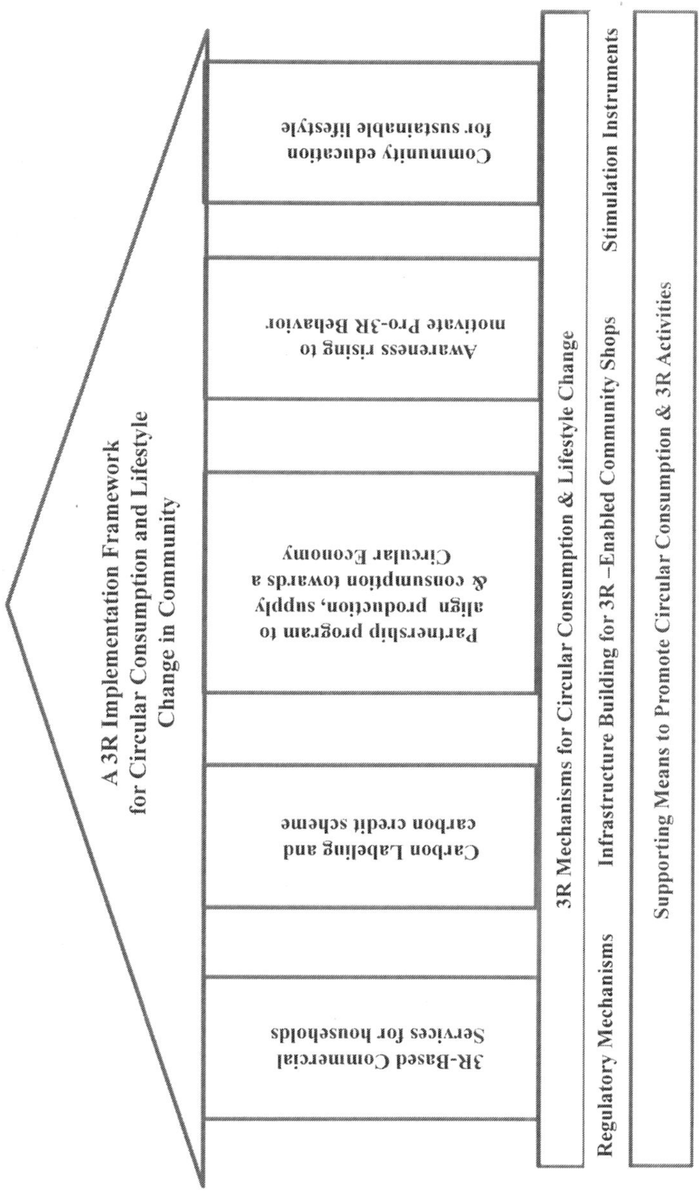

Fig. 11: A 3R Implementation Framework for Circular Consumption and Lifestyle Change in Community

The "reduction" principles of the 3R theory of citizens ' daily lives are embodying these goods in their own homes. A streamlined model of carbon quantification, interaction and rewards: carbon marking and carbon loan system. Carbon Labeling and Carbon Credit Scheme as shown in figure 12.

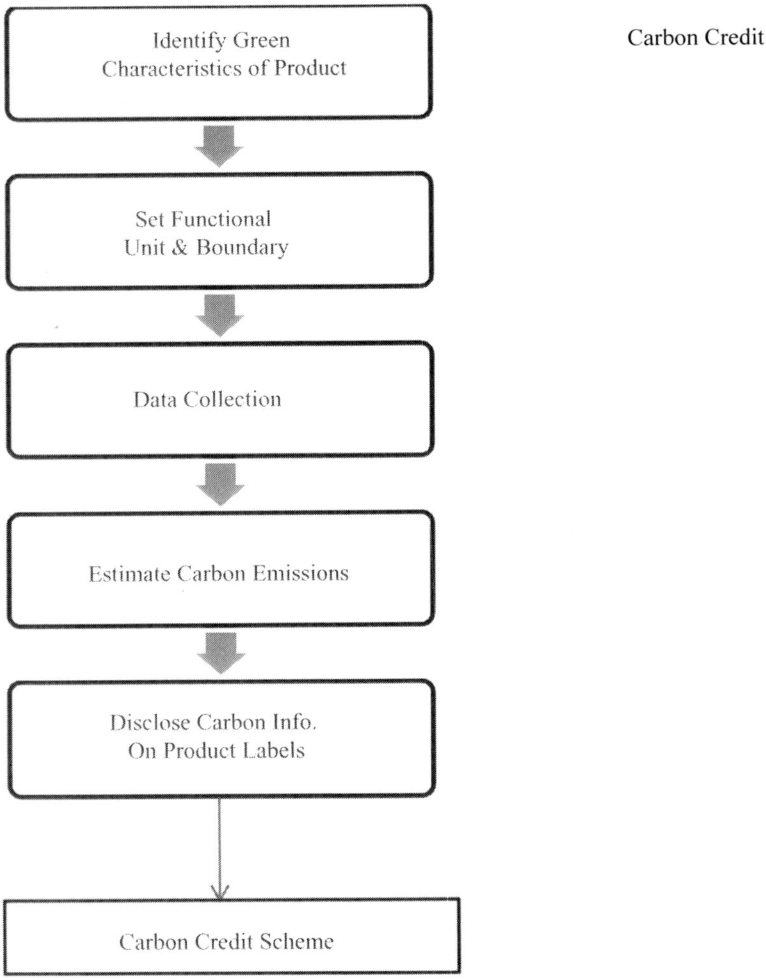

Fig. 12: Carbon Labeling and Carbon Credit Scheme

The new attribute of the label on carbon disclosure to low-carbon products is the consumer perception. The label makes it easier for consumers to compare products and leads to informed choices on low carbon commodities. The aim is the reduction of the environmental impact of consumption and changes the attitudes of residents to low- carbon products on 3R-shops. Each of the 3R-shop customer's 3R activities are converted and accumulated on the incentive card by using a 3R incentive card and an electronic information platform. Carbon credit flow is indicated in figure 13.

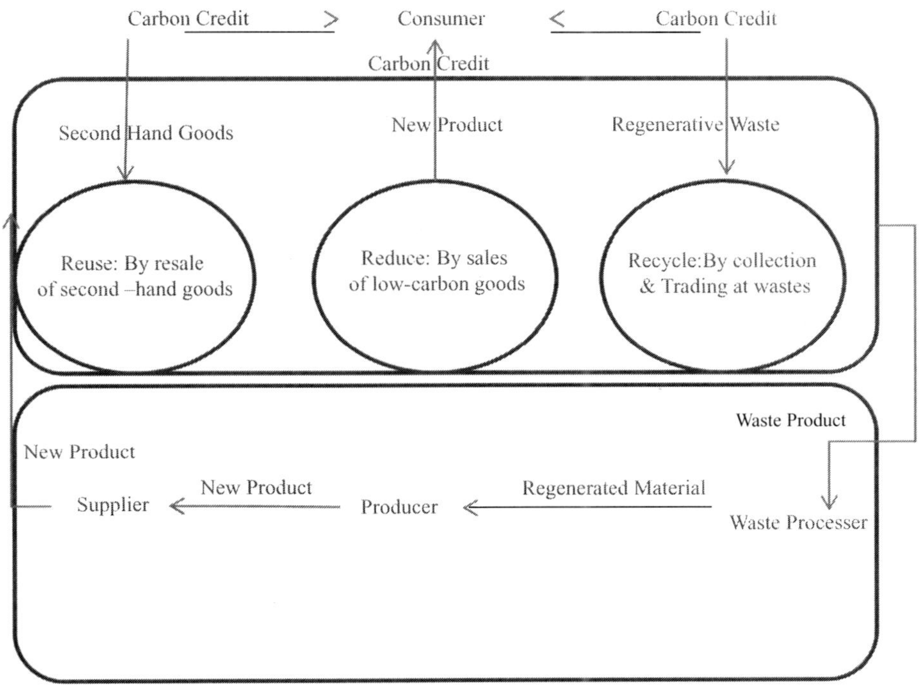

Fig. 13: Carbon Credit Flow

i. Awareness Rising to Motivate Pro-3R Behavior

- Conducted energy efficient products information campaigns, conservation of water.
- Bringing goods from low carbon and green households into towns and towns to allow residents to know and use the goods.
- To promote pro-3R behaviours, schools and organizations and the advantage of behavioral change.
- To inform customers in 3R-shops of the impact they have of their consumption on 3R and on the circular consumption agendas.

ii. Regulative instrument

CE and 3R. approaches have been adopted by Chinese government as new models for economic development and legislation has already been adopted.to support this strategy. Government funds are also available to foster the practice of 3R and CC by business, consumers and organizations. The 3R shops are located to serve local residents and has a business area of about 200 m (12 to 3 kilometers) and a service area of 5 to 6 localities. An incentive card and an associated electronic platform are the financial instruments for this framework.

A 3R framework has been developed to practice the concepts of circular consumption for residents of communities. This framework will support changes in patterns and lifestyles of consumption. The 3R-store companies have changed fundamentally the traditional way of retailing domestic products, reusing second-hand goods and collecting and recycling domestic waste in shop areas.3R services and advanced IT services were combined in 3R shops to produce, supply, consuming products in circular pattern system, supporting Circular Economy and supporting proactive 3R standard of life.21 These trends will probably go on, but there are still some major technological challenges, as well as economic or social behavioral problems related to the collection of recyclable waste and replacement of virgin material. Although implementation of 3R on a wide range is a far-reaching process in developing countries. The 3R should include environmental, social and economic plans, policies, strategies and programs in the national development agenda.

6.8 Conclusions

Decision variables can use the modeling applied in the region of waste management to address a specific issue, such as selecting the best waste management scheme and options. Models used as design decision-making aid for solid waste management are models based on LCA, CBA, and MCA. Decisions concerning municipal solid waste management are hard, as they are not only linked to financial and capital elements, but are also linked to social and environmental elements. Therefore, to offer a wide range of data to the decision-making process, it is essential to create and execute easy but reliable instruments as models are. Future research will concentrate on the evaluation of the acceptance of these 3R shops, the degree to which they have affected customer behavior and the use of circularity and sustainability actions.

6.9 Way Ahead

Models have been developed to analyze the externalities (Environment and social) of waste management, but that needs to bring up in the mindsets or consumer behavior. Non-monetary evaluation can be done in detail in respect to various operations of waste management. Geospatial techniques such as land suitability index can be done by considering the monetary as well as non-monetary terms of waste

References

Banar, M., Cokaygil, Z. and Ozkan, A. (2009) 'Life Cycle Assessment of Solid Waste Management Options for Eskisehir, Turkey', Waste Management. Elsevier Ltd, 29(1), pp. 54–62. doi: 10.1016/j.wasman.2007.12.006.

Behzadian, M. *et al.* (2010) 'PROMETHEE: A Comprehensive Literature Review on Methodologies and Applications', European Journal of Operational Research. Elsevier B.V., 200(1), pp. 198–215. doi: 10.1016/j.ejor.2009.01.021.

Bovea, M. D. and Powell, J. C. (2016) 'Developments in Life Cycle Assessment Applied to Evaluate the Environmental Performance of Construction and Demolition Wastes', Waste Management. Elsevier Ltd, 50, pp. 151–172. doi: 10.1016/j.wasman.2016.01.036.

Costi, P. *et al.* (2004) 'An Environmentally Sustainable Decision Model for Urban Solid Waste Management', 24, pp. 277–295. doi: 10.1016/S0956-053X(03)00126-0.

Garbie, I. H. (2015) 'An Analytical Technique to Model and Assess Sustainable Development Index in Manufacturing Enterprises', (August 2014). doi: 10.1080/00207543.2014.893066.

Hansen, T. L. *et al.* (2006) 'Life Cycle Modelling of Environmental Impacts of Application of Processed Organic Municipal Solid Waste on Agricultural Land (Easewaste)', Waste Management and Research, 24(2), pp. 153–166. doi: 10.1177/0734242X06063053.

Johnson, T. R. (2017) 'Municipal Solid Waste Management', in Routledge Handbook of Environmental Policy in China, pp. 302–313. doi: 10.4324/9781315736761. Journal of Scientific and Research Publications, 3(1), pp. 2250–3153. Available at: www.ijsrp.org.

Kapepula, K. M. *et al.* (2007) 'A Multiple Criteria Analysis for Household Solid Waste Management in the Urban Community of Dakar', Waste Management, 27(11), pp. 1690–1705. doi: 10.1016/j.wasman.2006.07.009.

Kumar, S. and Chakrabarti, T. (2010) 'Effective Municipal Solid Waste Management in India', (March). doi: 10.5772/8459.

Kumar, S., Negi, S., Mandpe, A., Singh, R.V., Hussain, A., 2018. Rapid composting techniques in Indian context and utilization of black soldier fly for enhanced decomposition of biodegradable wastes - A comprehensive review. J. Environ. Manage. 227, 189–199. doi:10.1016/j.jenvman.2018.08.096

Kumar, V. and Pandit, R. K. (2013) 'Problems of Solid Waste Management in Indian Cities', International

Mandpe, A., Kumar, S., Lakshmikanthan, P., Hettiarachchi, H., 2019a. Mining for recovery as an option for dumpsite rehabilitation: Case study from Nagpur, India. J. Environ. Eng. Sci. 15, 52–60. doi:10.1680/jenes.19.00021

Manyapu, V., Mandpe, A., Kumar, S., 2019. Corrigendum to "Synergistic effect of fly ash in in-vessel composting of biomass and kitchen waste" [Bioresour. Technol. 251 (2018) 114–120] (S0960852417321740)(10.1016/j.biortech.2017.12.039). Bioresour. Technol. 277, 225. doi:10.1016/j.biortech.2018.12.031

Morrissey, A. J. and Browne, J. (2004) 'Waste Management Models and their Application to Sustainable Waste Management', in Waste Management, pp. 297–308. doi: 10.1016/j.wasman.2003.09.005.

Reich, M. (2005) 'Economic Assessment of Municipal Waste Management Systems — Case Studies using a Combination of Life Cycle Assessment (LCA) and Life Cycle Costing (LCC)', 13, pp. 253–263. doi: 10.1016/j.jclepro.2004.02.015.

Samiha, B. (2013) 'The Importance of the 3R Principle of Municipal Solid Waste Management for Achieving Sustainable Development', 4(3), pp. 129–136. doi: 10.5901/mjss.2013.v4n3p129.

Verma, R. (2016) 'Ecotaxes : A Comparative Study of India and China'.

7

Novel Tactics for Sustainable Waste Management

Vivek Manyapu[a] and Ashootosh Mandpe[b,c,d*]

[a]*Department of Biotechnology, CSIR-Institute of Himalayan Bioresource Technology, Palampur, Himachal Pradesh, 176061, India*
[b]*Department of Civil Engineering, Indian Institute of Technology, Indore Indore 453 552, India*
[c]*Academy of Scientific and Innovative Research (AcSIR) Ghaziabad-201002 India*
[d]*CSIR-National Environmental Engineering Research Institute (CSIR-NEERI) Nehru Marg, Nagpur 440020, India*

ABSTRACT

Dealing with waste and its treatment is always a cumbersome process. Several researchers/industrialists from different countries are putting their valuable efforts to solve the issues arising due to ill-treated wastes and to recover energy and other valuable products from them. Earlier due to the huge burden of the waste, they were mainly disposed of in landfills, which pollute all three spheres viz. atmosphere, lithosphere, and hydrosphere. Therefore, traditional technologies need to be improvised which not only should be efficient in waste management but also should efficaciously recover energy from them. Along with improved modern landfill techniques, some of the improvised thermochemical, biochemical, and physicochemical techniques have been discussed in the chapter that includes incineration, gasification, pyrolysis, modern biomethanation, composting, and pelletization. In addition, the role of GIS in solid waste management has been discussed with a special focus on plastic waste in farmlands. GIS tools can be having a vital role in detecting the types of waste, quantifying the biomass production, spot the dumpsites, and finally creating a geo-database of the wastes generated.

Keywords: Bio-methanation; Composting; Gasification; GIS; Landfill; MSW

Acronyms

MSW - Municipal solid waste

GDP - Gross domestic product

SWM - Solid waste management

GHG - Greenhouse gas

MBT - Mechanical Biological Treatment

HC – Hydrocarbons

LFG - Landfill gas

LULC - Land use land cover

PWIS - Plastic waste indices

APW - Agricultural plastic waste

GIS - Geographic Information System

SSI - Site-sensitive index

DSS - Decision Support Systems

7.1. Developing Sector in Scientific and Technological Approach of Waste Management

From the very past, many traditional practices have been put into use for waste management like landfilling, composting, and recycling, which are practiced even today. However, the boundless upsurging waste compels the modern world to modify the traditional methods or to introduce innovative techniques with the grace of technology. The modern techniques can be based on biological, thermal or thermochemical methods. Biological treatments like anaerobic digestion and fermentation are commonly used while incineration, pyrolysis, and gasification come under thermal or thermochemical. The technologies are being evolved day-by-day for the treatment of municipal solid waste (MSW). Some peculiar technologies like direct casting, melting process, gasification, and fluidization are being used in some of the thermal power plants (Montejo et al. 2011). Novel technologies like gasification, pyrolysis, and plasma gasification, are being implemented for treating MSW, to produce some exorbitant products and heat energy (Brunner and Rechberger 2015).

The production of heat energy from waste is the *waste-to-energy* concept, which is nowadays is seeking more attention. The heat intensity is directly proportionate to the rate of absorption of the source of energy. The *waste-to-energy* can be followed by three processes:

a) Thermochemical process

i. *Combustion in excess of air:* It is the best technique to recover heat and produce electricity.

ii. *Gasification in very less air:* The waste is subjected to high temperature, and the gas is regained to use as fuel.

iii. *Pyrolysis in the total absence of air:* Steam is generated using the heat of treated waste which is processed at a very high temperature.

b) Biochemical process

The wet waste can be treated with anaerobic digestion to produce biogas. This biogas, when processed using gas engines, can be utilized to generate heat and electricity. The processed waste can be further utilized as biofertilizer/ manure. Using the biochemical process, valuable fuels like methanol, ethanol, biodiesel, methane, hydrogen, etc. can also be recovered (Pratap et al. 2021) Bioleaching is also a potential process in which a significant quantity of essential metals can be extracted from the slags, sludge, industrial wastes, batteries, etc. of different metal treatment plants/industries (Table.1).

Table 1: Bioleaching of metals by different microorganisms

Waste	Ash from a copper smelter	Copper converter slag	Zinc smelting slag	Industrial waste	Copper containing sludge	Spent lithium-ion batteries
Microorganisms	*Thiobacillus ferrooxidans* *Thiobacillus thiooxidans*	*Thiobacillus ferrooxidans* *Thiobacillus thiooxidans*	Moderately Thermophilic bacteria	*Penicillium simplicissimum*	*Sulfolobus acidocaldarius*	*Aspergillus niger*
Leaching Efficiency	Cu (87%)	Cu (86%) Ni (50%) Co (64%)	Al (82-84%) As (85-91%) Cu (86-88%) Zn (95-97%)	Al (75%)	Cu (99%)	Co (65%) Li (10%)
Leaching Condition	Shake flask=22 days pH=1.8 Ash content=50g/L	Shake flask=25 days pH=1.7 Ash content=50g/L	Shake flask pH=1.5 Smelting slag=50g/L	Shake flask Content=50g/L	The bioreactor was leached for 12 hours	Shake flask=18 days Content=100g/L

c) Physicochemical process

This method is to modify the physicochemical characteristics of solid waste. The combustible materials from the waste are converted into different sizes of pellets or refuse-derived fuel, which have high energy to generate steam and in turn electricity. The waste, before converting it into pellets, is properly dried, and separated from silt, sand, and grits by mechanical techniques. With high hydraulic force, the waste is converted into pellets.

For dry or low moisture-containing waste, a thermochemical process is used for the treatment. Whereas, the wet waste which contains organic matter is needed to be processed by the biochemical method (Moharir et al. 2019).

The innovative emerging techniques are compelling to think in a redefined way, changing the concept of waste from byproduct to cost reducer and performance improver (Osmani and Villoria-Sáez 2019).

"If you know how to handle waste, you may have GOLD with you."

7.2. Understanding the Principle of Available Waste Treatment Techniques and Exploring Its Novelty

7.2.1 Introduction

Among all the developing countries, India is the second-fastest-growing country in terms of economy as well as population. Our country has about 645 districts and approximately 7935 towns from which about 31.16 % of the Indian population i.e., 377 million (as per Census 2011) dwell in urban areas. It is forecasted that by 2026 the percentage of urban people will be 33.4 %. The proliferating population and simultaneous rise in the gross domestic product (GDP) result in the exaggerated generation of quantum of waste mainly from towns and cities. Annually, there is a very drastic boost in the generation of solid waste in Indian cities at the rate of 4.25 % from 6 million tons to 48 million tons within a span of 50 years from 1947 to 1997. The expected increase in waste generation in the cities is 0.3 billion by 2047. (CPCB, 1998) (Agarwal et al. 2015).

Expeditious growth of industrialization, urbanization, and population explosion in our country paved a way to shift a large amount of population from a rural to urban area which ultimately increases the solid waste generation in urban areas. With existing treatment techniques and measures for solid waste management (SWM), it is now becoming laborious to handle a tremendous amount of solid waste generated in a short period.

Along with the quantity of solid waste, now its quality has also become a stumbling block for treating and managing it. For example, one of the traditional methods of disposing of solid waste is landfilling (Mandpe et al. 2021a). However, with the huge quantity and variety of non-biodegradable waste mixed up with biodegradable one, it started creating problems like leachate, groundwater contamination, the atmospheric release of greenhouse gases (GHGs), and loss of resources/space. Along with these existing problems, there are indirect and barriers for example non-availability of large barren land for a landfill site, a non-segregated mixture of solid waste for collecting waste, the selection of long route because of traffic and nuisance for transportation, an increased variety of non-biodegradable waste for segregation and finally increased amount of GHG and demand of resources for disposal, etc. This ultimately demands advancements in traditional treatment techniques and sustainability for the betterment of humans as well as the environment.

To tackle these problems, we must know traditional disposal techniques of solid waste and further give advanced and sustainable solutions on the same. Following are the basic principles of SWM (Agarwal et al. 2015) :

1) The 5Rs System (Fig. 1)

i. Refuse = We should not buy goods or resources which are not necessary.

ii. Reduce = We should reduce the generation of garbage and we have to change our lifestyle which will aid in this approach.

iii. Reuse = We should try our best to reuse the things after proper treatment.

iv. Recycle = The recyclable things should be kept carefully and they should reach the recycling units or factories through *Kabadiwallahs* (rag pickers).

v. Recover = The waste can be treated with proper techniques to extract energy from the waste

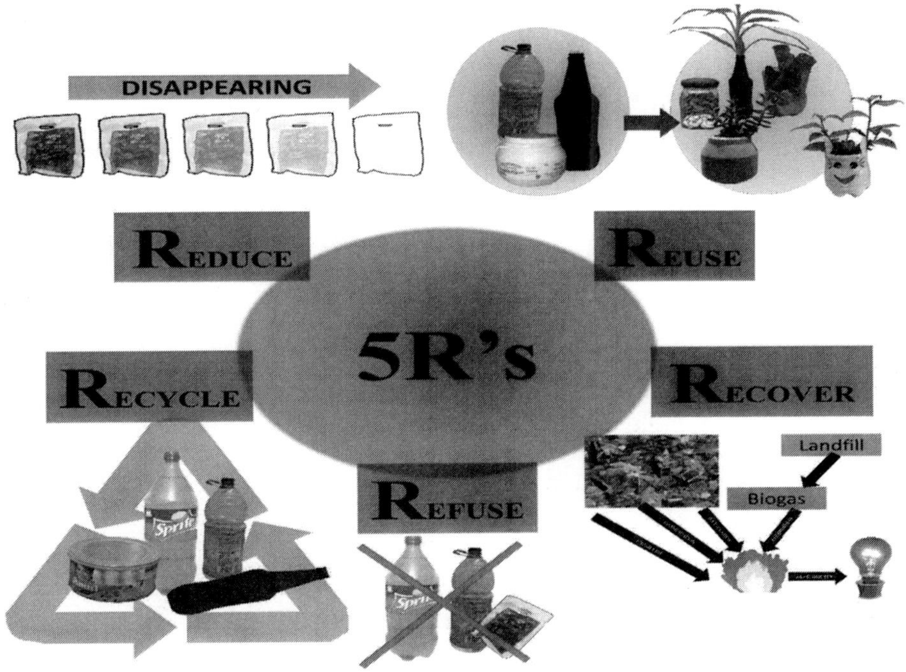

Fig. 1: The 5 R's system of waste management: The 5 R's namely Reduce, Reuse, Recycle, Refuse and Recover should be followed to manage the waste and it is highly effective in reducing the environmental deterioration due to waste disposal.

2) *Segregation at the Source itself:* There should be separate bins for separate wastes, for example, wet waste bins and dry waste bins; biodegradable waste bins and non-biodegradable waste bins; recyclable waste bins and non-recyclable waste bins; biodegradable waste bins, recyclable waste bins, and hazardous waste bins; etc. This approach reduces the cost of recycling with minimal labour cost.

3) *Treating different types of solid wastes differently:* Each type of waste must be treated in its specific way of treatment technique, e.g., the way the slaughterhouse waste is treated is different from that of general market waste.

4) *Treating the waste at the nearest possible place:* The way the solid waste must be treated should be a decentralized way. Possibly the waste should be treated at the source itself, starting from the houses (Agarwal et al. 2015).

Based on the above-mentioned principles, an ideal SWM for a rural place could be as under concerning technology and cost gradient (Table 2).

Table 2: Gradient of technology for solid waste management: major low to high technology options applicable to large-scale urban waste management

Technology	Unit Cost (per tonnes)	Energy Balance
Low to intermediate	Low to Intermediate	Negative to positive
Low to intermediate	Low to Intermediate	Negative to positive
High	High	1. Negative to positive
		2. MBT/aerobic: negative
		3. MBT/anaerobic: positive
		4. Anaerobic digestion: positive
		5. Incineration: positive (highest)

Note: MBT=Mechanical Biological Treatment.
Source (Bogner et al. 2008)

7.2.2 Modern Landfill

Urban areas in India gather about 70-90% of MSW generated, whereas smaller cities and towns contribute less than 50% of waste generated. More than 91% of the MSW collected formally is landfilled in open lands and dumps (Bharti et al. 2017). The landfill is the most straightforward and widely used treatment technique for MSW because it is the most simple and cheap technology. However, open burning of MSW and landfill fires together releases 22,000 tons of pollutants into the lower atmosphere. The pollutants include carbon monoxide (CO), carcinogenic hydrocarbons (includes dioxins and furans), particulate matter, nitrogen oxides (NO), and sulphur dioxide (SO_2) (Bharti et al. 2017). Landfill gas which is the combination of CH_4 and CO_2 is known as biogas (Saleem et al. 2016). This biogas can be utilized efficiently for domestic purposes, as a vehicle fuel, and for generating heat and electricity (Fig. 2).

On a global basis it was found that the landfill gas (LFG) emission increased to 68 Mt/year in 2017 from 57 Mt/year in 2000 (Saunois et al. 2019). In India, every year, from the landfills the methane emission is estimated to be about 16 million metric CO_2 equivalents (International Energy Agency, 2008) (Table 3).

According to Planning Commission Report (2014) annually 62 million tons of MSW is produced in cities that can generate 439 MW of electricity from combustible materials and 72 MW from LFG and 5.4 million metric tons of compost for agriculture use. CH_4 has 23 times higher global warming potential than CO_2. The methane gas from landfills is having an essential role in energy production since it gets converted into carbon dioxide and water (Joshi and Ahmed 2016). These gas emission rates are highly differing due to hydrological, geotechnical, and geological factors of the environment, which have environmental impacts. The biotic and abiotic factors lead to the generation of gas at the landfill, which is the combination of methane and carbon dioxide, which is called biogas (Saleem et al. 2016).

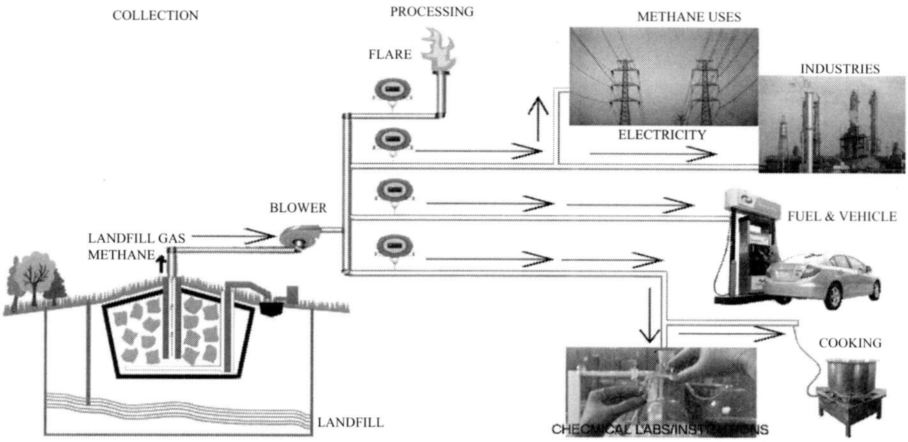

Fig. 2: Landfill producing and utilizing GHG-Methane (CH_4): carefully designed landfills can be used to collect methane gas generated due to anaerobic digestion of organic components. After proper processing, this methane can be put into the efficient purpose of generating electricity, heat for industries, fuel for vehicles, for cooking, for chemical testing and reactions, etc. Apart from the gas, the leachate percolating down can also be used as liquid fertilizer.

Table 3: The energy potential from the landfill gas available at selected sites of India

Sl.No.	Regions	City	Energy potential
1.	Balswa, Gazipur and Okhla	Delhi	8.4 MW
2.	Deonar and Gorai	Mumbai	5.6 MW
3.	Pirana	Ahmadabad	1.3 MW
4.	Urli	Pune	0.7 MW

Chemical oxidation of water and metals like Aluminium gives rise to abiotic gas. Leachate produced by Aluminium produces hydrogen gas through a redox reaction. Metallic aluminium hydration and bottom ash result in gas production. There may be a hazard of LFG or the gradual leakage will result in global warming if there is mismanagement (Saleem et al. 2016). This LFG-CH_4 is being used for various purposes

1. To fuel industrial boilers;
2. To generate electricity using IC engines, gas or steam turbines;
3. To produce a substitute natural gas after removal of CO_2 and trace components.

Although the range of electrical output is from a small 30 kW to 50 MW through microturbines and steam turbine generators respectively, most of the plants are in the range of 1–15 MW (Bogner et al. 2008). Many places in India have installed and many are planning to install generators to extract energy from the LFG-CH_4 (Table 4).

Table 4: Waste-to-energy installed and planned projects in different places under the Government of India

Sl. No.	Place	Capacity (in MW)	Installed/ Planned projects	References
1.	Solapur, Maharashtra	3	Installed	Ministry of New and Renewable Energy (2014)
2.	Okhla, Delhi	16	Installed	
3.	Bangalore	8	Planned	
4.	Hyderabad	11	Planned	
5.	Pune	10	Planned	
6.	Ghazipur	12	Planned	
7.	Narela	24	Planned	(Joshi and Ahmed 2016)

7.2.3. Modern Thermochemical processes of waste management

7.2.3.1. Incineration

In Tokyo's central part, there are about 19 waste incineration plants. Many people consider waste incineration plants negatively because of their dirtiness and air pollution. But the modern technology is proving it wrong and it is more efficient and eco-friendly (Winter 2015).

According to Japanese technology, the conventional stoker furnace is a potentially efficacious technology that can generate electricity from the heat produced during the processing of the waste and it is very much efficient to control greenhouse gases (Japan Environmental Sanitation Center 2012) (Fig. 3).

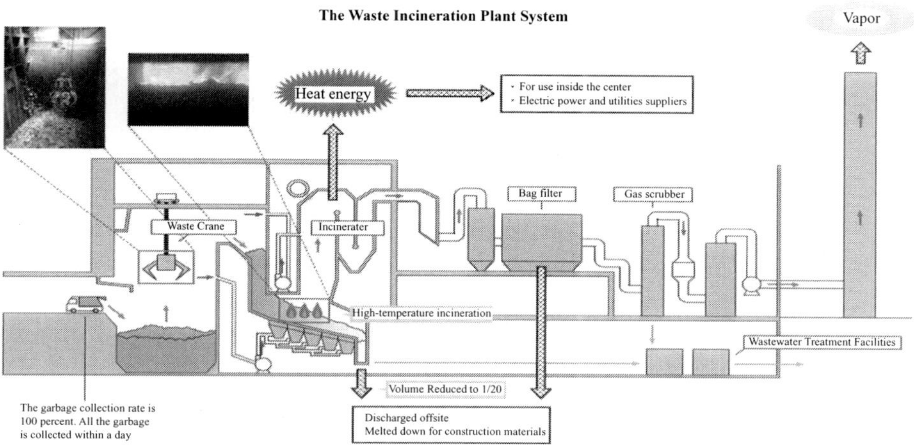

Fig. 3: Modern incineration plant: Incineration helps in reducing the huge amount of waste within a short period and simultaneously generate high calories of heat energy which can be used for generating electricity, insulating the houses/buildings, etc. and the resultant products (gas, ash, and water) are released into the environment after **undergoing proper treatments.**

The Japanese manufacturers have characterized the following features in the modern incinerators (Japan Environmental Sanitation Center 2012):

- Improved combustion quality
  - Helps in lowering combustion air ratio from 2 to 1.4 or 1.3
  - Helps in resolving uneven burns by infusing high-temperature air
  - Helps in improving mixing combustion gas and air
  - Helps in improving oxygen-enriched combustion and combustion quality, which results in clean incinerated ash
- Improved quantification and control of
  - Supply of garbage
  - Stoker cooling both water-and air-cooling, water cooling, and air cooling
  - Improved durability of fire-resistant: Tile cover for radiation heat transfer surface on boiler
  - furnace wall material
  - Incinerate ash recycling
  - Operation control system sophisticating.

The leader in the construction and management of incinerators is Japan. With its excellent standard of technology, Japan treats the waste through and disposes of low-calorie to high-calorie garbage. Apart from Japan, many other successful countries are making efficient use of the heat generated from the incineration (Table 5).

Table 5: Modern incineration plants designed by Japanese technology situated in different parts of the world

Industry	Place	Capacity	Remarks
Mitsubishi Heavy Industries, Environmental & Chemical Engineering Co., Ltd.	Singapore	4,320t/day	The largest scale plant
JFE Engineering Corporation	Thailand	100t/day.	Industrial waste incinerator
Hitachi Zosen Corporation	Taipei	--	Most of the garbage generated in the country is being disposed of and reduced in a sanitary manner.
Takuma Co., Ltd.	Beijing	1,600t/day	An electric generating facility with the capacity of 30,000 KW and steam condition of 4MPa and 400°C

Source: (Winter 2015)

7.2.3.2. Gasification

The gasification of waste is nowadays attracting more and more attention due to its unique process in which the waste is converted into gaseous fuel which can be used for various purposes. It is estimated that 80% of the chemical energy present in the waste can be utilized in the form of gaseous fuel. For this reason, this process is more efficient and capable of reducing waste than incineration (Breeze 2018).

Gasification is the partial oxidation process. It can be taken up in the presence of air/oxygen where the waste is partially burnt to derive the required amount of energy. Japan is the popular use of this technology, where it is significantly reducing the enormous amount of waste. This process is very popular in Japan as it is widely used to curb the volume of waste. By the use of oxygen aided high-temperature gasification, any inorganic material within the waste is melted and converted into inert slag, that can be used as construction material. In Japan, this specific slag conversion process is called slagging gasification. A process known as pressure-swing adsorption, that produces the oxygen in a particular plant, is used for the slagging gasifier.

During gasification, syngas is produced. After proper cleaning of the gas, it can be used for various industrial purposes. Usually, it is burnt to produce electricity through boiling the water to convert it into steam and passing through steam turbines. Instead, it is more efficient if it is run through a gas turbine or gas engine. Whatever the use of Syngas, the gas must be cleaned scrupulously to exclude any other reactive gases which may damage the engines. Irrespective of the type of plant, the flue gases emerging from the engines or the boiler should be cleaned to regulate the emissions of NOx, CO, Suspended PM, and any volatile organic compounds (Breeze 2018).

7.2.3.3. Pyrolysis

Pyrolysis is defined as a thermal process of disintegration in an anoxic environment persisting in which the biomass can be transformed into different forms of by-products, like solid (biochar, tar), gas (syngas), and liquid (bio-oil). It can be sub-categorized into slow pyrolysis, fast pyrolysis, flash pyrolysis, vacuum pyrolysis, intermediate pyrolysis, and hydro-pyrolysis based on the operational conditions (Bridgwater and Peacocke 2000; Jones et al. 2009; Laird et al. 2009; Onay and Kockar 2003; Tripathi et al. 2016; Vamvuka 2011) Slow pyrolysis is also known as the conventional pyrolysis method. Slow pyrolysis is considered a conventional process in the production of charcoal as it yields maximum biochar produce when compared with fast pyrolysis and gasification (Raheem et al. 2015). The process of slow pyrolysis is considered the most viable process for producing biochar of higher quality and grade which is a consistent and reliable product for its application in agricultural activities (Song and Guo 2012) Initially, the pre-pyrolysis happens followed by the solid disintegration conforming to the main pyrolysis process at a higher rate which tends to the formation of pyrolysis products. Lastly, the char decomposes at an extremely low rate, and thus the carbon-rich solid residue namely biochar is formed (Suganya et al. 2016). Fig. 4. Shows how the syngas produced during the process of pyrolysis is used for various purposes through generating energy.

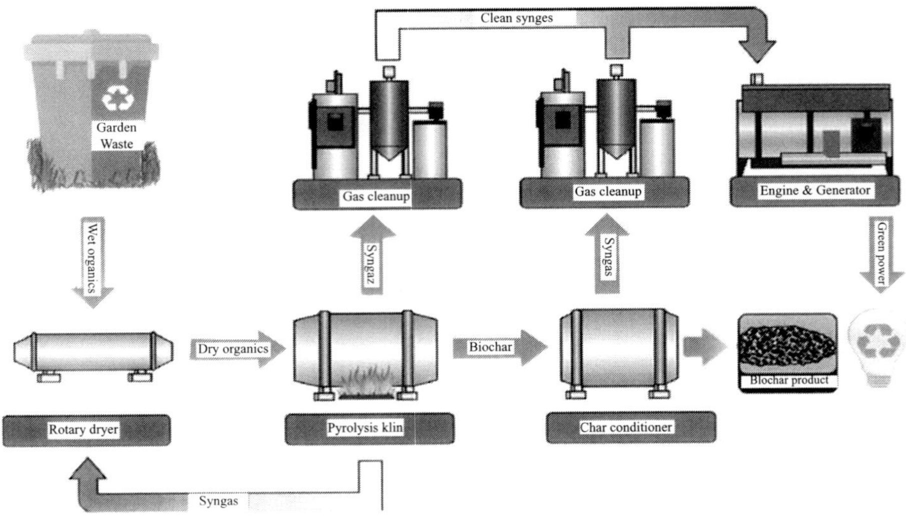

Fig. 4: Syngas production from the waste: Syngas is one of the significant gas generated during the process of pyrolysis. This syngas is efficaciously helpful in generating electricity and also the residual syngas can be used for the pyrolysis process which requires a huge amount of heat energy.

7.2.4. Modern Biochemical Processes of Waste Management

7.2.4.1. Biomethanation

An agrarian country like India is producing immense agricultural waste, which is having the capability of producing heat through combustion. When the waste is in wet condition, the efficiency of combustion reduces. Despite understanding the value of organic waste and the economic benefits, the municipal solid waste that mostly comprises of organic waste is dumped into landfills. Improper disposal in the landfills results in different environmental problems, like odor emission, attracting vermins, different obnoxious/toxic gases release, contamination of groundwater, and waste of valuable land and biomass (Shin et al. 2001)called MUlti-step Sequential batch Two-phase Anaerobic Composting (MUSTAC called Multi-step Sequential Batch Two-phase Anaerobic Composting (MUSTAC). Further, open landfilling causes various detrimental issues like greenhouse gas emission, soil and groundwater pollution, foul smell, stray animals, rodents, vectors like flies and mosquitoes, and many more respiratory disorders (Idris et al. 2004). Technological advancements of digester design will be highly effective in recovering methane from wet waste that can result in good outcomes for the users, society and even helps the environment reduce the waste issues. The outcome may be energy in the form of heat, light, electricity; production of high-quality organic fertilizer, improved sanitation through less of high-quality organic fertilizer, improved sanitation through less pathogenic influences, employment generation, reduced workload for rural women, especially for firewood collection and cooking, traditional chullahs can be replaced with biogas stoves and many more environmental benefits like reduced pollution and the greenhouse effect. Therefore, biogas technology can be upgraded that help in contributing to conservation and national development.

7.2.4.2. Composting Technology

In the earlier periods, composting was mainly carried out in heaps or pits outdoors in open fields *viz.* windrow system, static pile system passively aerated windrow system, forced aerated static pile system, vermicomposting (Manyapu et al. 2017), Bangalore method, Indoor method, Coimbatore method, NADEP method, etc. Nevertheless, the trend is now changed into more rapid and more efficient composting technology. In-vessel composting is being used to treat a huge amount of organic waste to produce nutrient-enriched compost (Manyapu et al. 2018). The advantages of in-vessel composting are many, which include moderating the internal temperature, moisture, and thus providing a favourable microclimate that helps the microorganisms' population to proliferate rapidly. Apart from the design, researchers have worked on different combinations of microbial consortia, the addition of different additives (Kumar et al. 2018; Mandpe et al. 2021b; Mandpe et al. 2021c; Mandpe et al. 2020; Mandpe et al. 2019a; Manyapu et al. 2018; Paliya et al. 2019; Singh et al. 2019a; Singh et al. 2019b)solid waste management (SWM and enzymes which help in enhancing the process (Mandpe et al. 2019b; Negi et al. 2020). Moreover, pre-treatment of the substrate makes it readily available for decomposition (Manyapu et al. 2018).

Nowadays, the agriculture system is converting from a chemical-free system to an organic farming system. Therefore, the marketing of compost will be very much profitable.

7.2.5. Physicochemical Technology

7.2.5.1. Pelletization

Simply pelletization refers to the process of compressing the biomass waste into pellets that can be used as fuel. Pellets are cylindrical with 6-25 mm diameter and 3-50 mm in length (Alakangas 2010). The bulk density of the biomass gets increased to around 700 kg/m^3 (Sokhansanj and Turhollow 2004)\nparticle density, and particle size. Biomass densified into bales and cubes increases its bulk density to 120 to 500 kg/m3. The \nhandling properties of pellets and cubes are similar to those properties for grains. The existing, well-developed, conventional \nbulk handling equipment can be used to handle pelleted and cubed biomass. Although cubed biomass is easier and safer to \nhandle and store, it is more expensive as a feedstock for a conversion plant than biomass bales. The objective of this research \nwas to develop a base cost for cubing using the existing technology. The cost of corn stover bales delivered to a biorefinery \n64 km (40 miles. Moreover, they are made into homogeneous shapes and structures which will be very helpful for automated feeding into boiler systems. The rise in the cost of fuel and wood is pressurizing alternative methods. The agro-industrial organic waste is highly preferred to replace fuel and wood. The raw materials, which are used for pellet production, are fibrous residues from croplands and the food processing industry, *viz.* straw, husk, pulps, etc. The different types of biomass densification are bailing, briquetting, extrusion, and pelletization (Tumuluru et al. 2010). Pelletizing includes different stages *viz.* raw material pre-treatment, pelletization, and post-treatment. In the pre-treatment process, the size of the raw materials is reduced followed by drying and conditioning. Pelletization is running the raw material through the machinery to pressurize the raw material into pellets. After that, the pellets are cooled and screened to eliminate small particles (Holm et al. 2011; Holm et al. 2007; Holm et al. 2006; Stelte et al. 2011)

7.3. Make Innovation a Weapon for Our Arsenal

7.3.1 Implementation of Geographic Information System in Solid Waste Management

Geographic Information System (GIS) is becoming a boon and prominent tool in nearly every sector of the world because of its high data handling and processing ability, less use of resources, more accuracy level along time efficiency. Integrated GIS deals with spatial and non-spatial data for giving output in the form of images and maps, which further can be used for decision-making, running queries, etc. The amalgamation of these integrated GIS applications with advanced and sustainable solid waste management systems will give prodigious and phenomenal solutions to the problems we are facing with MSW. The Italian and European researchers implemented a web-based GIS technique to optimize the waste collection and source separation that is used to recycle was as efficient as 80% (Rada et al. 2013).

As GIS provides data like land use land cover (LULC) of existing Earth surface, streets, landmarks, water bodies, geotechnical data of soil, statistical data of population, etc. which plays an important role mainly in collection and transportation of MSW. Following are some applications of GIS in MSW management in India studied by experts:

One simple procedure adopted for site selection of landfill in the Municipality of Nikea, Athens is explained below (Chalkias and Lasaridi 2009)most of it being spent on salaries and fuel. It is therefore crucial to improve the WC&T system through routing optimisation. Geographic Information System (GIS:

Step 1: Identification of problem and selection of study area

The Municipality of Nikea, which is one of the largest system in the Attica Region, situated in the South-West part of the Athens metropolitan area. It is focusing on the identification of the optimum scenario for waste collection of the mixed waste, for minimizing the collection time, distance traveled, human-efforts, and consequently financial and environmental costs.

Step 2: Data collection

Data must be collected regarding the study area like soil type, LULC, groundwater source, population, streets, commercial and public facilities, roads, railways, settlement and its type, and quantity of solid waste generated, distance traveled by existing waste transportation systems, its capacity, route, etc.

Step 3: Reallocation of waste collection bins

Routing optimization for shortest possible distance, minimum time, least fuel consumption, and gas emissions by using ArcGIS Network Analyst GIS software.

Step 4: Routing – Network analysis

For modifying vehicle routes, modified Dijkstra's algorithm was used. By the use of active data structures like d-heaps in the context of ArcGIS network analyst commercial GIS software, the algorithm was improvised (Chalkias and Lasaridi 2009)most of it being spent on salaries and fuel. It is therefore crucial to improve the WC&T system through routing optimisation. Geographic Information System (GIS. Table 6 elaborates the use of different GIS and Decision Support Systems (DSS) techniques in different regions for different purposes.

Table 6: Chronological order of DSS and GIS in Municipal Solid Waste Management in India
Source (Deswal and Laura 2014)

S.N.	Author	Techniques	Area of Application	Study location
1	Natesan and Suresh (2002)	GIS-based decision support system, MCE models: AHP, FIC, ANN and Delphi	Landfill site selection	Chennai
2	Asadi et al. (2007)	ArcGIS platform	Data Management, Validation and Analysis module. Through these modules transfer station capacity, site-sensitive index (SSI) and suitable site for the landfill was selected.	Medak
3	Reddy and Padmaja (2007)	GIS-based DSS (Decision Support System)	Optimal Route Analysis for transportation of waste.	Hyderabad
4	Ohri and Singh (2010)numerical models and expert opinions in multi objective decision making scenario. Decision Support Systems (DSS	ArcGIS and NET Engine programming tools.	Estimation of waste generation, onsite operation, primary collection, transportation, resource recovery, and disposal.	Varanasi city
5		Satellite data and Linear regression analysis.		Nagpur
6		Sustainability-based decision support system		Punjab
	Katpatal and Rama Rao (2011)	Satellite data and Linear regression analysis.		Nagpur

7.3.1.1. The agricultural plastic waste (APW) on the land was analyzed and georeferenced using the GIS database. The methodology was carried out by the following:

a. Creating a set of plastic waste indices (PWIs) for every crop and its plastic use;

b. Preparing land use map in ArcMap;

c. Selecting and demarcating the crops producing plastic waste on the land use map
d. Detecting of plastic materials utilized for protecting the crop on the map (film and net);
e. Attributing the PWIs for every crop and plastic use to the various characteristics in the land;
f. Quantitatively evaluating and geo-referring of APW;
g. Creating geo-database briefing the full data on the APW;
h. Realizing the APW maps;
i. Identifying suitable locations of the APW collection centers.

For the survey of the land, a field computer with a Global Navigation Satellite System receiver was used along with a Pro 6H receiver and a Juno 5 handheld computer, with the help of Trimble Terra Sync and global positioning system (GPS) Pathfinder Office software (Blanco et al. 2018) Through GIS techniques, the plastic waste estimation for different croplands has been shown in Table 7.

Table 7: Different types of plastic wastes generated from different croplands

Type of waste	Crop type	Quantity of plastic waste (kg/ha/year)
Pipes for irrigation	1. Olive trees 2. Greenhouses	50 104
Covering films and nets	1. Vineyards with nets only 2. Vineyards with films and nets	159 773
Plastic bags	1. Olive trees 2. Cereals	0.5 2.7
Plastic containers	1. Olive trees/cereals 2. Vineyards	0.6 4.0

7.3.1.2. In Europe, GIS-MCDA can be used to spot suitable areas for placing some biomass provisions to promote renewable energy. The study of different areas has been carried out using the criteria like environmental, economic, physical, social, etc., and using two modeling functions like AHP and WLC (Jeong and Ramírez-Gómez 2017) (Table 8).

Table 8: Different criteria used by GIS-MCDA

Criteria		Evaluation factors	
1.	Physical	a) Slope b) Orientation	c) Land use d) Lithology
2.	Environmental	a) Crop type b) Biomass resources	c) Vegetation type d) Ecosystem
3.	Economic	a) Site access b) Transport cost	c) Possible demand d) Economic development
4.	Social	a) Population b) Visibility	c) Settlement d) Tourism

7.4. Conclusion

It is always a very difficult process to dispose waste when the quantity is very high, yet there are some ways that can be implemented to reduce the burden of waste on the environment. Biological/biochemical/physicochemical methods should be of top priority because they are in harmony with nature. However, they are only confined to organic waste. Therefore, higher preference is given to thermochemical methods like incineration, gasification, and pyrolysis. These thermochemical processes irrespective of the type of waste reduce the quantity and generate energy or fuel if handled properly. Landfilling should not be encouraged as a waste management process. Even though it is maintained with due care because the risk of pollution will always persist either in the form of methane catching fire or leachate leakage into soil and groundwater.

References

Agarwal, R., Chaudhary, M., Singh, J., 2015. Waste management initiatives in India for human well-being 23.

Alakangas, E., 2010. New European pellets standard, in: Eubionet-III. Finland, p. 9.

Asadi, S.S., Vuppala, P., Reddy, M.A., 2007. Remote sensing and GIS techniques for evaluation of groundwater quality in municipal corporation of Hyderabad (Zone-V), India. Int. J. Environ. Res. Public. Health 4, 45–52. https://doi.org/10.3390/ijerph2007010008

Bharti, V., Singh, J., Singh, A., 2017. A Review on Solid Waste Management Methods and Practices in India. Trends Biosci. 10, 4065–4067.

Blanco, I., Loisi, R.V., Sica, C., Schettini, E., Vox, G., 2018. Agricultural plastic waste mapping using GIS. A case study in Italy. Resour. Conserv. Recycl. 137, 229–242. https://doi.org/10.1016/j.resconrec.2018.06.008

Bogner, J., Pipatti, R., Hashimoto, S., Diaz, C., Mareckova, K., Diaz, L., Kjeldsen, P., Monni, S., Faaij, A., Qingxian Gao, Tianzhu Zhang, Mohammed Abdelrafie Ahmed, Sutamihardja, R.T.M., Gregory, R., 2008. Mitigation of global greenhouse gas emissions from waste: conclusions and strategies from the Intergovernmental Panel on Climate Change (IPCC) Fourth Assessment Report. Working Group III (Mitigation). Waste Manag. Res. 26, 11–32. https://doi.org/10.1177/0734242X07088433

Breeze, P., 2018. Advanced Waste-to-Energy Technologies, in: Energy from Waste. Elsevier, pp. 65–75. https://doi.org/10.1016/B978-0-08-101042-6.00007-8

Bridgwater, A.V., Peacocke, G.V.C., 2000. Fast pyrolysis processes for biomass. Renew. Sustain. Energy Rev. 4, 1–73. https://doi.org/10.1016/S1364-0321(99)00007-6

Brunner, P.H., Rechberger, H., 2015. Waste to energy – key element for sustainable waste management. Waste Manag., Special Thematic Issue: Waste-to-Energy Processes and Technologies 37, 3–12. https://doi.org/10.1016/j.wasman.2014.02.003

Chalkias, C., Lasaridi, K., 2009. A GIS based model for the optimisation of municipal solid waste collection: The case study of Nikea, Athens, Greece. WSEAS Trans. Environ. Dev. 1.

Deswal, M., Laura, J.S., 2014. Application of GIS in MSW management in India. Int. J. Eng. Res. Dev. 10, 24–32.

Holm, J.K., Henriksen, U.B., Hustad, J.E., Sørensen, L.H., 2006. Toward an Understanding of Controlling Parameters in Softwood and hardwood Pellets Production. Energy Fuels 20, 2686–2694. https://doi.org/10.1021/ef0503360

Holm, J.K., Henriksen, U.B., Wand, K., Hustad, J.E., Posselt, D., 2007. Experimental Verification of Novel Pellet Model Using a Single Pelleter Unit. Energy Fuels 21, 2446–2449. https://doi.org/10.1021/ef070156l

Holm, J.K., Stelte, W., Posselt, D., Ahrenfeldt, J., Henriksen, U.B., 2011. Optimization of a Multiparameter Model for Biomass Pelletization to Investigate Temperature Dependence and to Facilitate Fast Testing of Pelletization Behavior. Energy Fuels 25, 3706–3711. https://doi.org/10.1021/ef2005628

Idris, A., Inanc, B., Hassan, M.N., 2004. Overview of waste disposal and landfills/dumps in Asian countries. J. Mater. Cycles Waste Manag. 6, 104–110. https://doi.org/10.1007/s10163-004-0117-y

Isaksson, L., Hugelius, G., Ishizawa, M., Ito, A., Janssens-Maenhout, G., Jensen, K.M., Joos, F., Kleinen, T., Krummel, P.B., Langenfelds, R.L., Laruelle, G.G., Liu, L., Machida, T., Maksyutov, S., McDonald, K.C., McNorton, J., Miller, P.A., Melton, J.R., Morino, I., Müller, J., Murgia-Flores, F., Naik, V., Niwa, Y., Noce, S., O'Doherty, S., Parker, R.J., Peng, C., Peng, S., Peters, G.P., Prigent, C., Prinn, R., Ramonet, M., Regnier, P., Riley, W.J., Rosentreter, J.A., Segers, A., Simpson, I.J., Shi, H., Smith, S.J., Steele, L.P., Thornton, B.F., Tian, H., Tohjima, Y., Tubiello, F.N., Tsuruta, A., Viovy, N., Voulgarakis, A., Weber, T.S., van Weele, M., van der Werf, G.R., Weiss, R.F., Worthy, D., Wunch, D., Yin, Y., Yoshida, Y., Zhang, W., Zhang, Z., Zhao, Y., Zheng, B., Zhu, Qing, Zhu, Qiuan, Zhuang, Q., 2019. The Global Methane Budget 2000–2017 (preprint). Atmosphere – Atmospheric Chemistry and Physics. https://doi.org/10.5194/essd-2019-128

Japan Environmental Sanitation Center, 2012. Solid Waste Management and Recycling Technology of Japan - Toward a Sustainable Society.

Jeong, J.S., Ramírez-Gómez, Á., 2017. Renewable energy management to identify suitable biomass facility location with GIS-based assessment for sustainable environment. Energy Procedia 136, 139–144. https://doi.org/10.1016/j.egypro.2017.10.310

John, S., 2010. Sustainability-based Decision-Support System for solid waste management. Int. J. Environ. Waste Manag. 6, 41–50. https://doi.org/10.1504/IJEWM.2010.033982

Jones, S.B., Valkenburt, C., Walton, C.W., Elliott, D.C., Holladay, J.E., Stevens, D.J., Kinchin, C., Czernik, S., 2009. Production of Gasoline and Diesel from Biomass via Fast Pyrolysis, Hydrotreating and Hydrocracking: A Design Case (No. PNNL-18284 Rev. 1). Pacific Northwest National Lab. (PNNL), Richland, WA (United States). https://doi.org/10.2172/950728

Joshi, R., Ahmed, S., 2016. Status and challenges of municipal solid waste management in India: A review. Cogent Environ. Sci. 2. https://doi.org/10.1080/23311843.2016.1139434

Katpatal, Y.B., Rama Rao, B.V.S., 2011. Urban Spatial Decision Support System for Municipal Solid Waste Management of Nagpur Urban Area Using High-Resolution Satellite Data and Geographic Information System. J. Urban Plan. Dev. 137, 65–76. https://doi.org/10.1061/(ASCE)UP.1943-5444.0000043

Kumar, S., Negi, S., Mandpe, A., Singh, R.V., Hussain, A., 2018. Rapid composting techniques in Indian context and utilization of black soldier fly for enhanced decomposition of biodegradable wastes - A comprehensive review. J. Environ. Manage. 227, 189–199. https://doi.org/10.1016/j.jenvman.2018.08.096

Laird, D.A., Brown, R.C., Amonette, J.E., Lehmann, J., 2009. Review of the pyrolysis platform for coproducing bio-oil and biochar. Biofuels Bioprod. Biorefining 3, 547–562. https://doi.org/10.1002/bbb.169

Mandpe, A., Kumari, S., Kumar, S., 2020. Composting: A Sustainable Route for Processing of Biodegradable Waste in India, in: Hettiarachchi, H., Caucci, S., Schwärzel, K. (Eds.), Organic Waste Composting through Nexus Thinking: Practices, Policies, and Trends. Springer International Publishing, Cham, pp. 39–60. https://doi.org/10.1007/978-3-030-36283-6_3

Mandpe, A., Paliya, S., Kumar, S., Kumar, R., 2019a. Fly ash as an additive for enhancing microbial and enzymatic activities in in-vessel composting of organic wastes. Bioresour. Technol. 293, 122047. https://doi.org/10.1016/j.biortech.2019.122047

Mandpe, A., Paliya, S., Kumar, S., Kumar, R., 2019b. Fly ash as an additive for enhancing microbial and enzymatic activities in in-vessel composting of organic wastes. Bioresour. Technol. 293, 122047. https://doi.org/10.1016/j.biortech.2019.122047

Mandpe, A., Pratap, V., Lakshmikanthan, P., 2021a. Chapter 9 - Landfill mining, in: Kumar, S., Kumar, R., Pandey, A. (Eds.), Current Developments in Biotechnology and Bioengineering. Elsevier, pp. 197–229. https://doi.org/10.1016/B978-0-12-821009-3.00010-5

Mandpe, A., Tyagi, L., Paliya, S., Chaudhry, S., Motghare, A., Kumar, S., 2021b. Rapid-in-house composting of organic solid wastes with fly ash supplementation: Performance evaluation at thermophilic exposures. Bioresour. Technol. 337, 125386. https://doi.org/10.1016/j.biortech.2021.125386

Mandpe, A., Yadav, N., Paliya, S., Tyagi, L., Ram Yadav, B., Singh, L., Kumar, S., Kumar, R., 2021c. Exploring the synergic effect of fly ash and garbage enzymes on biotransformation of organic wastes in in-vessel composting system. Bioresour. Technol. 322, 124557. https://doi.org/10.1016/j.biortech.2020.124557

Manyapu, V., Mandpe, A., Kumar, S., 2018. Synergistic effect of fly ash in in-vessel composting of biomass and kitchen waste. Bioresour. Technol. 251, 114–120. https://doi.org/10.1016/j.biortech.2017.12.039

Manyapu, V., Shukla, S., Kumar, S., Rajendra, K., 2017. In-vessel composting: a rapid technology for conversion of biowaste into compost. Open Access Int. J. Sci. Eng. 2, 6.

Ministry of New and Renewable Energy, 2014. Annual Report 2014-15. New Delhi.

Moharir, R.V., Gautam, P., Kumar, S., 2019. Chapter 4 - Waste Treatment Processes/Technologies for Energy Recovery, in: Kumar, S., Kumar, R., Pandey, A. (Eds.), Current Developments in Biotechnology and Bioengineering. Elsevier, pp. 53–77. https://doi.org/10.1016/B978-0-444-64083-3.00004-X

Montejo, C., Costa, C., Ramos, P., Márquez, M. del C., 2011. Analysis and comparison of municipal solid waste and reject fraction as fuels for incineration plants. Appl. Therm. Eng., Selected Papers from the 13th Conference on Process Integration, Modelling and Optimisation for Energy Saving and Pollution Reduction 31, 2135–2140. https://doi.org/10.1016/j.applthermaleng.2011.03.041

Natesan, U., Suresh, E.S.M., 2002. Site suitability evaluation for locating sanitary landfills using GIS. J. Indian Soc. Remote Sens. 30, 261–264. https://doi.org/10.1007/BF03000369

Negi, S., Mandpe, A., Hussain, A., Kumar, S., 2020. Collegial effect of maggots larvae and garbage enzyme in rapid composting of food waste with wheat straw or biomass waste. J. Clean. Prod. 258, 120854. https://doi.org/10.1016/j.jclepro.2020.120854

Ohri, A., Singh, P., 2010. Development of decision support system for municipal solid waste management in India: a review. Int. J. Environ. Sci. 1, 440–453.

Onay, O., Kockar, O.M., 2003. Slow, fast and flash pyrolysis of rapeseed. Renew. Energy 28, 2417–2433. https://doi.org/10.1016/S0960-1481(03)00137-X

Osmani, M., Villoria-Sáez, P., 2019. Chapter 19 - Current and Emerging Construction Waste Management Status, Trends and Approaches, in: Letcher, T.M., Vallero, D.A. (Eds.), Waste (Second Edition). Academic Press, pp. 365–380. https://doi.org/10.1016/B978-0-12-815060-3.00019-0

Paliya, S., Mandpe, A., Kumar, S., Kumar, M.S., 2019. Enhanced nodulation and higher germination using sludge ash as a carrier for biofertilizer production. J. Environ. Manage. 250, 109523. https://doi.org/10.1016/j.jenvman.2019.109523

Pratap, V., Bombaywala, S., Mandpe, A., Khan, S.U., 2021. Chapter 12 - Solid Waste Treatment: Technological Advancements and Challenges, in: Karri, R.R., Ravindran, G., Dehghani, M.H. (Eds.), Soft Computing Techniques in Solid Waste and Wastewater Management. Elsevier, pp. 215–231. https://doi.org/10.1016/B978-0-12-824463-0.00014-8

Rada, E.C., Ragazzi, M., Fedrizzi, P., 2013. Web-GIS oriented systems viability for municipal solid waste selective collection optimization in developed and transient economies. Waste Manag. 33, 785–792. https://doi.org/10.1016/j.wasman.2013.01.002

Raheem, A., Wan Azlina, W.A.K.G., Taufiq Yap, Y.H., Danquah, M.K., Harun, R., 2015. Thermochemical conversion of microalgal biomass for biofuel production. Renew. Sustain. Energy Rev. 49, 990–999. https://doi.org/10.1016/j.rser.2015.04.186

Reddy, M., Padmaja, V., 2007. Development of GIS based decision support tool for optimal route analysis for transportation of solid waste. Presented at the Proceedings of 22nd International Conference on Solid Waste Technology and Management Philadelphia, PA USA, March, p. 2007.

Saleem, W., Zulfiqar, A., Tahir, M., Asif, F., Yaqub, G., 2016. Latest technologies of municipal solid waste management in developed and developing countries: A review. Int. J. Adv. Sci. Res. 1, 22–29.

Saunois, M., Stavert, A.R., Poulter, B., Bousquet, P., Canadell, J.G., Jackson, R.B., Raymond, P.A., Dlugokencky, E.J., Houweling, S., Patra, P.K., Ciais, P., Arora, V.K., Bastviken, D., Bergamaschi, P., Blake, D.R., Brailsford, G., Bruhwiler, L., Carlson, K.M., Carrol, M., Castaldi, S., Chandra, N., Crevoisier, C., Crill, P.M., Covey, K., Curry, C.L., Etiope, G., Frankenberg, C., Gedney, N., Hegglin, M.I., Höglund-

Shin, H.S., Han, S.K., Song, Y.C., Lee, C.Y., 2001. Multi-step sequential batch two-phase anaerobic composting of food waste. Environ. Technol. 22, 271–279. https://doi.org/10.1080/09593332208618283

Singh, P., Ghosh, D., Manyapu, V., Yadav, M., Majumder, S., 2019. Synergistic impact of Iron (III) oxide nanoparticles and organic waste on growth and development of Solanum lycopersicum plants: new paradigm in nano-biofertilizer. Plant Arch. 19, 339–344.

Sokhansanj, S., Turhollow, A., 2004. Biomass Densification - Cubing Operations and Costs for Corn Stover. https://doi.org/10.13031/2013.16480

Song, W., Guo, M., 2012. Quality variations of poultry litter biochar generated at different pyrolysis temperatures. J. Anal. Appl. Pyrolysis 94, 138–145. https://doi.org/10.1016/j.jaap.2011.11.018

Stelte, W., Holm, J.K., Sanadi, A.R., Barsberg, S., Ahrenfeldt, J., Henriksen, U.B., 2011. Fuel pellets from biomass: The importance of the pelletizing pressure and its dependency on the processing conditions. Fuel 90, 3285–3290. https://doi.org/10.1016/j.fuel.2011.05.011

Suganya, T., Varman, M., Masjuki, H.H., Renganathan, S., 2016. Macroalgae and microalgae as a potential source for commercial applications along with biofuels production: A biorefinery approach. Renew. Sustain. Energy Rev. 55, 909–941. https://doi.org/10.1016/j.rser.2015.11.026

Tripathi, M., Sahu, J.N., Ganesan, P., 2016. Effect of process parameters on production of biochar from biomass waste through pyrolysis: A review. Renew. Sustain. Energy Rev. 55, 467–481. https://doi.org/10.1016/j.rser.2015.10.122

Tumuluru, J.S., Wright, C.T., Kenney, K.L., Hess, R., 2010. A Technical Review on Biomass Processing: Densification, Preprocessing, Modeling and Optimization. Am. Soc. Agric. Biol. Eng. St Joseph Mich. https://doi.org/10.13031/2013.29874

Vamvuka, D., 2011. Bio-oil, solid and gaseous biofuels from biomass pyrolysis processes—An overview. Int. J. Energy Res. 35, 835–862. https://doi.org/10.1002/er.1804

Winter, T., 2015. Advanced Waste Disposal Technology Makes Tokyo the Cleanest City / JapanGov - The Government of Japan. Tokyo.

8

An Innovative Approach Towards Sustainable Development

***Rucha Vikas Moharir*[1,2,3] *and Sunil Kumar*[1,3]**

[1]*Academy of Scientific and Innovative Research, CSIR-National Environmental Engineering Research Institute Campus, Nehru Marg Nagpur-440 020, India*
[2]*Department of Civil Engineering, Priyadarshini Indira Gandhi College of Engineering, Digdoh Hills, Nagpur-440 019, India*
[3]*CSIR- National Environmental Engineering Research Institute (CSIR-NEERI), Nehru Marg, Nagpur-440020, India*

ABSTRACT

Increasing burden on the waste stream is a matter of grave concern resulting in quest of emerging techniques and modifications to the existing treatments and methods to be more aligned in a way that it should not harm the surrounding environment and maintain the ecological balance. Current chapter based on the available treatment options to be adapted for reaching the concept of sustainability. Various techniques are elaborated in this chapter including bio-mining concept, remediation ways, waste-to-wealth approach and use of waste plastic in making of brooms. Changing lifestyle patterns and increasing load on waste end are somewhere following the parallel path and hence improvisations are necessary to tackle an escalating problem of waste management. Main objective of this chapter is to emphasize the available treatment options focusing on waste utilisation in an environmentally sound manner.

Keywords: Solid waste, bio-mining, waste to wealth approach, remediation techniques, composting of waste, refuse derived fuel, biogas production, waste management, sustainability.

8. Introduction

The idea of sustainable development stands on the concept of maintaining the best possible environment within the availability of minimal resources and for the convenience of the society. Now-a-day the term “sustainability” gained a lot of attention due to challenges and problems faced in tackling the current situation of global pollution. One of the vital areas of research needs to be addressed in concern with sustainable development is the new innovations to enhance sustainability (Silvestre and Silva, 2014). Innovation is the main driving factor which circuitously deals with the day to day fluctuating environment and changing life style patterns of human society (Huisingh et al., 2013) hence more and more attention is now grabbed by this new area of research which becomes a key element for organisations, institutions and various regions to attain a systematic approach towards sustainable development (Silvestre, 2015a). The best possible route to attain sustainability passes through an innovative approach and leads to maintain a good social and ecological environment. Past literature also supports this area of upcoming challenge where 'sustainable' word used for the cause of benefit of the entire ecosystem (Silvestre, 2015b). However in today's technological world, somewhere we are neglecting the basic phenomenon to maintain a balanced situation against each and every used technological intervention for striving best towards sustainability by the institutions, organisations and other sectors but unfortunately it seems to be driving only in the direction of technology but not towards sustainability (Almeida et al., 2013, Bharti et al., 2019; Wainaina et al., 2020inoculum size of 3 mL, temperature of 30 °C, 150 ppm, and time of 5 days). Due to variety of applications of each technology in every individual sector, many new ways are introduced for the better enhancement of the existing versions but the need of the hour is to introspect all in terms for maintaining environmental balance. Proper check must be their like either our innovations are helpful in reducing carbon footprint or else we are only contributing innovations at the cost of our environment. For this cause, more and more environmentally friendly ways should be approached to tackle such vital issues on environmental management.

8.1 Bio-mining

8.1.1 Introduction

Bio-mining is a technique used to extract metals from the ores, rock course or even to clean the sites loaded with metal via microbial routes. Microorganisms play a very crucial part of extraction of these metals from the object. It is one of the environmental friendly ways to resolve the problem faced in metal extraction techniques. The function of microorganisms works on the principal of secreting the organic compound from the metals via chelation and then providing these compounds to the cells for further fulfilling their requirement of electrons. This was one of the greatest discoveries that microorganisms secrete metals for their survival happened in the year 1990s. Bio-mining over mining has so many advantages including the by-products generated after bio-mining viz. metabolites or gases are not harmful to the ecosystem whereas the pollutants generated after mining are well destroyers of the environment. Bioremediation works on the same principle where the by-products generated are environmental friendly and harmless. Microbes can be immunized in the environment having these metals, ores and oils. They have the capacity of converting these carbon-based compounds into usable products for generating energy in the cell and ultimately the reason for their survival. Microbes do this job so efficiently and effectively which for human beings is a page of mystery. The importance of these biological treatments relies in the final output of the treatment in terms of end products which are bio friendly.

8.1.2 Recent Techniques of Remediation

Mining is a very old technique used for extracting metals from ores. As a consequence this technique has many adverse impacts on the environment. So, for environmental protection and to safeguard our lands and water resources, government environmental laws enforced that the mining should be achieved by ecological and sustainable ways. For the instance, rehabilitation and remediation come in context to fit in the concept of ecological restoration of mining lands (Perez-Lopez et al., 2010). Out of all the acid mining drainage techniques developed till now, very few are cost effective and most of all are very expensive technological solutions. Sustainability is one of the factors which need to be maintained while applying these techniques all alone whereas some of the studies recommended that to convert the mining lands into wetlands rather than converting it into drylands (Otte and Jacob, 2008). Some of the useful techniques are illustrated in Fig. 1 and described in this segment.

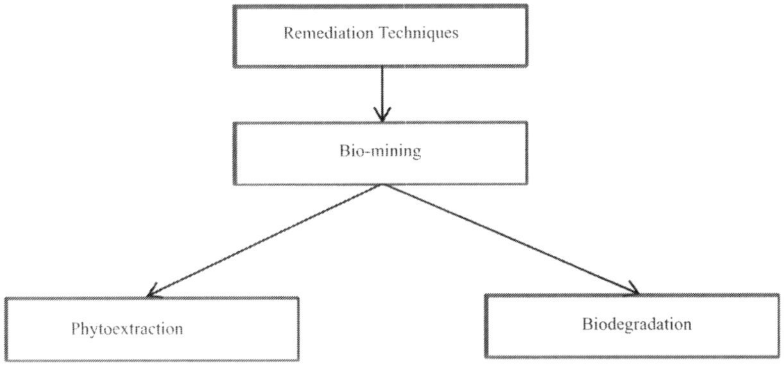

Fig. 1: Remediation techniques to overcome the problem faced in mining activities

8.1.2.1 Bioremediation

In mining process, land used for extraction of ores then just be filled with toxic metals and is no more available for any biological purpose as it contains many difficulties including soil contamination. Coming out of this menace, bioremediation of the wasteland is one of the ways reducing soil sickness with the help of such microbes which are capable of extracting metals from the soil and making soil usable for further use (Mandpe et al., 2019b). This metal extraction method is performed by two ways either in situ or ex situ, though ex situ is equally in use but in situ is mostly preferred as the cost involvements are less in case of this method in comparison to ex situ method (Azubuike et al., 2016).

8.1.2.1.1 Phytoextraction

Phytoextraction is also a remediation technique mainly used to remove some harmful elements and compounds mostly heavy metals and toxic substances from soil and water by means of plants. Plants roots absorbs water near the roots subsequently traces some metals and other harmful compounds helping soil to restoration. As these plants accumulates high amount of toxic elements including large number of heavy metals called as hyper-accumulators. Some plants may have low accumulation capacity reducing lower rate of toxics helping the system to get rid of the soil contamination problem. An overview of the process is schematically represented in Fig. 2.

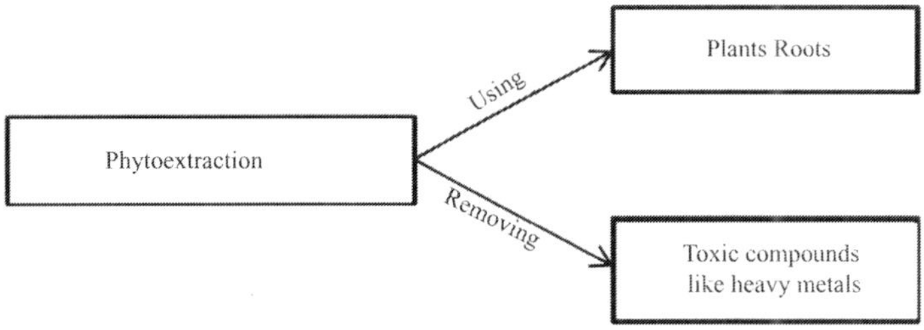

Fig. 2. Overview of phytoextraction process

8.1.2.1.2 Biodegradation

Biodegradation technique is used to degrade the material biologically with the help of bacterial or fungi capable of degrading the targeted substrate helps in maintaining ecological balance in a sustainable manner (Fig 3.). Mechanism works on the principle of bio-deterioration, then bio-fragmentation and then bio-assimilation. Microbes first deteriorate the material then convert it into more simple compounds for their uptake and then start degrading the material containing carbonaceous material works as a source of energy to them (Lucas et al., 2008).

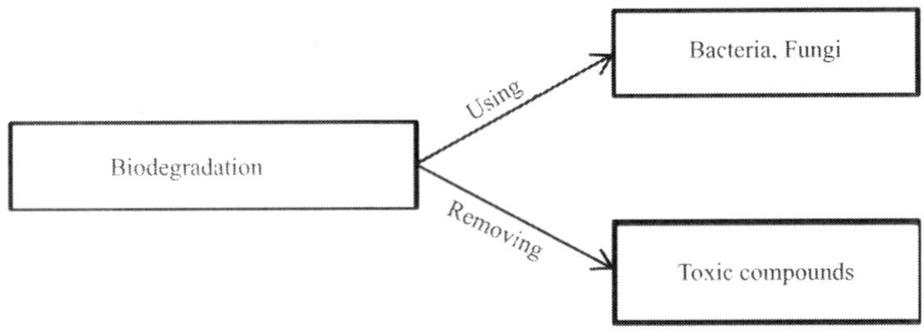

Fig. 3: Overview of biodegradation process

8.1.3. Impact of biochar on phytostabilization

In many ecosystem media there are several organic and inorganic contaminants due to mining activities (Karbowska, 2016). In mining activities many toxic elements can be leached out in natural environments including soil which is a cause of concern as it also affects the natural ecosystem surviving on the same like when enters in ground surface polluting ground water (He et al., 2005). This extraction process ends with complete excavation of the existing land and left unattended many times after its use without any treatment so landed as a vanished zone (Gibbs and Salmon, 2015). Many factors gets affected due to mining comprising of physical, chemical as well as biological factors exacerbating water and soil contamination, environmental degradation, biodiversity loss and health risks to humans (Confalonieri et al., 2014). Excavated soil from mining is generally lack of minerals, nutrients, biologically deficient, cohesion less, containing low organic materials and with high heavy metal content in short not fit in to any further use from environmental viewpoint (Gonzalez et al., 2017). Various factors are in line which affects heavy metal toxicity and mobility in soils like their metal properties, binding state, soil properties, environmental factors including organic matter content, pH, type of soil, texture of soil, redox conditions, and root exudates acting as chelates (Rodriguez et al., 2009). Proper consideration should be given to these factors working in favour of efficient remediation strategies when it comes to wasteland of mining for better developmental factors and enhanced ecological response.

8.1.4 Way to Lead

Bio-mining has many applications including remediation, phytoextraction and biodegradation to combat the problem challenged in mining activities. This approach stands on the grounds of implementing newer technologies accomplishing higher metal yields. Bioleaching from metal ores is one of the effective methods works on the principle of metal extraction via biological means. Improvisation in addition to newer techniques will lead to solve the complex issues escalating in mining process.

8.2 Wealth Generation from Waste

Growing population, rising demands and increasing production rate will finally result in forming new challenges and questions against how to tackle the amount of waste generated by the system. There is a growing need of managing the generated waste properly by adapting system friendly techniques and environmental friendly routes to maintain the sustainability within the system. If waste is not tackled feasibly may responsible to cause many adverse impacts considering ecological solvability. Treatments should be well aligned with the waste management strategies and more focus should be given on reduction of waste generation and treatment at the source options. 3R principle should be followed strictly wherever it is possible to combat over a situation. Considering waste as a valuable asset will help resolve the problems more efficiently and developing such waste-to-energy concepts is the need of an hour. Emphasis should be on maximum utilisation of waste in a way of reusing or reproducing things to overawe the increasing waste peak all over the globe (Moharir et al., 2019).

8.2.1 Waste to Energy (WtE) generation

Utilisation of waste in a form of energy is a very good alternative to manage the waste in an efficient way (Zhao et al., 2016). WtE concept is based on targeting such waste which can no longer be further recycled or reused so that it can be utilised in producing some energy which can be further absorbed in any treatment wherever it is required in that form. Various treatments are available and improvised in this aspect to produce energy from municipal solid waste to be utilised further such as incineration but this process showed some adverse impacts in the form of emitting some harmful gases threatening the environment. Taking into account the adversities caused, new alternatives came in front with system needed improvisation with better treatment facilities reducing the environmental threat. WtE generation is a way of reducing load on waste management streams converting waste into a valuable product and generating economy out of waste is a good covenant. Traditional methods for waste treatments have some loopholes to be look after for better replacement options if possible like in case of landfilling (Nagendran et al., 2016). Landfilling of waste comes with many drawbacks but is a very old method of treating waste. Though modern landfills are far way better and improvised still selecting this option is a last priority to elect (Mandpe et al., 2019a).

8.2.1.1 Importance of WtE Concept

Waste is either disposed in an open land or landfilled without any prior treatment else ends up floating in natural river bodies polluting lands and causing water pollution. In quest of other alternatives waste to energy concept came in front producing a form of energy from generated waste before disposing it anywhere decreasing the menace (European Commission, 2010). Application of WtE concept help in combating many drawbacks by converting waste into fuel or energy generation from waste is a better solution to be used in many treatments where these forms of energy can be utilised. Fuel generated from waste can also replace fossil fuels reducing the stress on the system and mitigating adverse impacts caused by these treatments like in combustion and incineration process many harmful gases are released in environment being a major source of greenhouse gas emission.

Several methods are implemented for producing energy from waste and that energy is further utilised for various purposes. Waste to energy conversion methods includes anaerobic digestion, incineration, gasification and gas recovery from landfills.

8.2.2 Composting of Waste

Composting of waste is one of the oldest methods of treating organic waste. Only organic waste products are considered in this method including kitchen waste containing organic matter. Microorganisms then turned the collected organic waste material into humus like substance which then be used as a compost/fertiliser for soil (European Commission, 2015). This method is globally accepted being environment friendly method to utilise the waste in the form of manure. Advantages of composting includes richness in soil quality, improving soil texture, strengthening plant roots, refining soil fertility, conditioning of soil, maintaining environmental balance using less chemicals in treatment. Many factors work for effective conversion of waste to manure like temperature, moisture content, oxygen content, organic content, carbon to nitrogen ratio and type of compost. Composting is performed in many ways and also practiced in different categories of which some are highlighted in this segment (Ahn et al., 2011; Kumar et al., 2018).

8.2.2.1 Windrow Composting

Windrow composting is further categorised in different types considering the type of waste which is to be treated including untreated waste, size reduced waste and pulverised waste. Each category is processed differently demanding specific treatments according to waste type. These categories are well explained in the following segment.

8.2.2.1.1 Windrowing of Untreated Waste

This process is performed with least requirements. This type of waste is directly sent to treatment zone where further paving is carried out. Manual and mechanical turning are performed depending on the need and watering is mostly done using hose. After sufficient turnings, decomposed waste is then transferred to storage are followed by removing unwanted material from the waste. This windrowing is only performed with the waste having less particle size.

8.2.2.1.2 Windrowing Performed with Size Reduced Waste

Mostly preferable method as size reduction and then elevation is much easier than just disposing the waste without reducing its size. Conveyor belts are used in case of transferring the waste from one place to another and size reduction is done with the help of hammer mills further size reduced waste is then transported through conveyor belts or sometimes with the help of tractors. Waste unloading is done mostly through machines or by front-end loaders.

8.2.2.1.3 Windrowing Method for Partly Fermented Waste

This method is a combination of above methods with advanced mechanism as large rotating drum is installed instead of hammer mill for reducing the waste size. By using this method, around 100 tonnes of waste can be treated having a retention time of 6-7days. As this system mechanism is slightly different, waste reduction is attained by the abrasion caused between the waste materials and rotating drum with varying hardness. Rotation helps for proper mixing of waste and also to achieve homogeneity. To maintain aerobic conditions, air is passed inside the rotating drum and moisture content is adjusted as per the requirements. Vent pipe is attached to the system for removing the gases generated inside the drum as aerobic conditions takes place for minimizing the bad odour.

This method has all the features; only instead of the provision of hammer mill a large rotating drum is installed. Hundred tons of waste can be treated which has a retention time for nearly up to 6-7 days. In this method size reduction of the waste is achieved by abrasion when waste material rubbed together in the rotating drum with differing hardness. Further due to rotating movement, waste is properly mixed and attains homogeneity. Moisture content is adjusted and then air is passed in drum to maintain aerobic conditions. Generated gases are then removed through vent pipe to minimize odour.

8.2.2.2 In-vessel Composting

In vessel composting is a very popular method of composting. As name suggests, in this method composting is practiced in vessel, container or suitable equipment (Robert, 1992). This method works on principle of bioreactor, and air-temperature flow is maintained throughout vessel which is either made of plastic or concrete. Aerobic condition is maintained and the factors responsible for smooth functioning of the system are monitored throughout the process (Mandpe et al., 2021a). Air pipe, temperature probe, pH probe is provided to monitor the inner position within the vessel which allows to maintain optimum conditions for aerobic decompositions. This technique is mostly preferred for municipal waste containing organic matter (Manyapu et al., 2019).

8.2.2.3 Tunnel Composting

This type of composting is completely modular, enclosed and controlled type of system where air is passed through air pipes and a controlled environment is created throughout the tunnel to maintain healthy aerobic environment to receive good quality compost.

8.2.2.4 Vermicomposting

Vermicomposting is a very effective, environmentally friendly technique to convert waste into manure. In this technique, worms are introduced in the waste material which alters entire waste into manure. Mostly introduced worms are red wigglers, white worms and earthworms. As worms are introduced for conversion, there are many benefits of this method such as soil fertility, soil conditioning, restoration of plants, improves soil texture, enhances water holding capacity (Yavitt, 2015).

8.2.3 Refuse Derived Fuel (RDF)

8.2.3.1 Introduction

The name itself suggests that it's a fuel derived from waste mainly municipal solid waste, commercial waste or industrial waste. Waste material having calorific value can be utilised in cement kilns, sometimes can be a replacement of fossil fuels meeting the specifications. RDF is specifically a waste containing combustible material of non-recyclable products like plastics, cardboard, paper and other discarded items. Screening is specifically done to separate each material by proper sorting. Sorting includes separation of components like glass, stones, metals, ferrous and non-ferrous material and then shredding is performed to homogenize the whole waste material or sometimes pellets are made to further use in replacement of fossil fuels or used as fuel in cement kilns (Rao et al., 2017).

8.2.3.2 Conversion process of biofuels

All over the world, biofuels are mainly considered as liquid fuels including bioethanol, biomethanol and biodiesel but biopower and biogas are also considered as biofuels. Out of all the biofuels, bioethanol and biodiesel are the most utilised biofuels having largest share in the global market. Bioethanol produced from sugarcane in Brazil and from corn in the United States raised the market with around 71% of global bioethanol production capacity (Gaur et al., 1998). Food based bioethanol are mainly from sugarcane, corn and biodiesel generally made from fats, vegetable oil and plants. Process of biofuel formation starts with production of biomass, then harvesting takes place, after that sorting is performed and then transportation to biorefineries is the final step involved diagrammatically represented in Fig. 4. In biorefinery, biomass is converted to biofuel, bioproducts, heat and power and then it is made available to the users. 30% moisture containing mixture termed as biomass whereas dry mass is termed for mixture having 10% moisture content. Biofuels are now steeping towards grabbing most of the bio-market and being a platform for discussion for sustainable resources. Any biomass feedstock can be converted to biofuels adapting the conversion routes of chemical, thermochemical, biochemical or hydrothermal pathway.

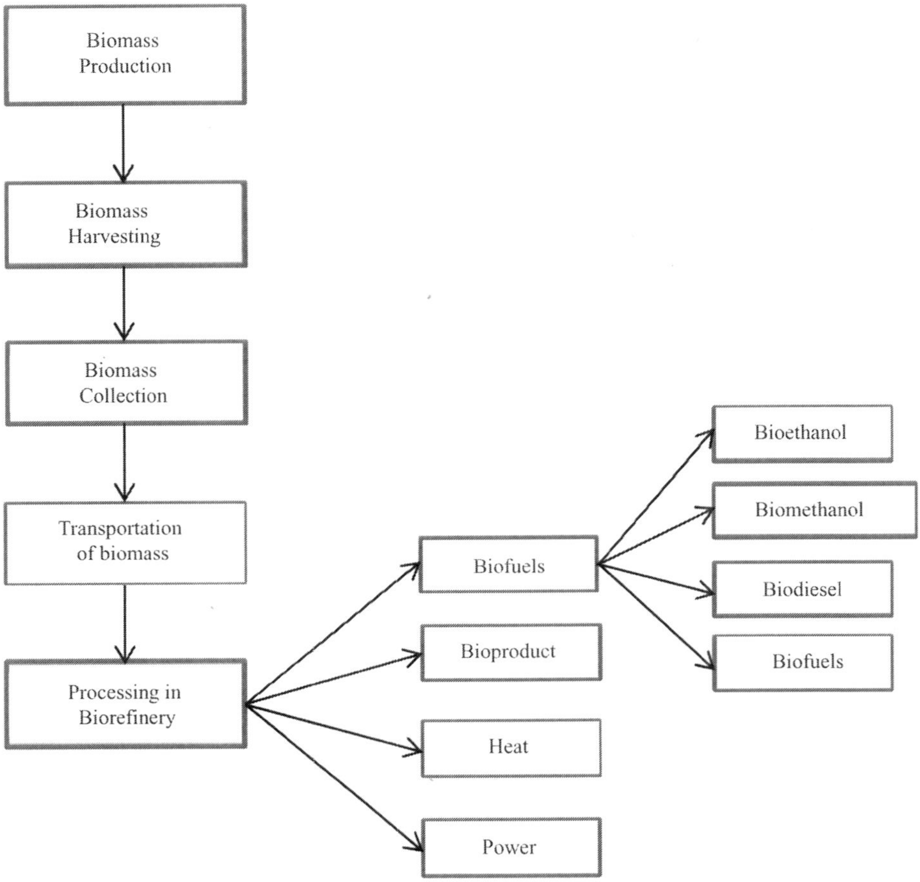

Fig. 4: Schematic representation of biofuel production

8.2.3.3 Biodiesel Production

Seed crops contain high protein level with oil in certain amount containing carbon atoms. Biodiesel can be made from food and non-food oils, waste cooking oils and animal fats. The quality of biodiesel completely depends on the type of oil, conversion process and the operating conditions. Two largest oilseed crops are mainly palm and soybean. Protein reach food produces the meal with high economic values. Other biodiesel feedstock's contains small amount of vegetable oil and animal fats such as canola, camelina, jatropha. Characteristics of soybean are represented in Table 1.

Table 1: Contents of soybean oil

Name	Protein	Oil	Carbohy-drate	Main Component in oil	Density (at 20 °C)	Melting Point	Heat of combustion
Soybean	41%	21%	29%	Triacylglycerol	0.916g/ml	0.6 °C	9.0 kcalg-1

8.2.3.3.1 Biodiesel from plant oil

Palm oil, soybean oil and rapeseed oil are the main sources of producing biodiesel. Soybean oil is the main source of production of biodiesel in United Sates. Small amount of vegetable feedstock is also present in animal fats such as canola, camelina and jatropha. Soybean is sold as a highly nutritious animal feedstuff amongst all the sources available to produce biodiesel. Biodiesel mainly produced from soybean and sunflower oil seeds are an excellent feedstock for carbonate the ultimate source of biodiesel (Yasar, 2018).

8.2.4 Biogas Production

Biogas mainly comprises of hydrocarbons which are in combustible form and able to produce heat and energy from waste material like food waste. Biological waste is first converted to some useful end products like biogas by the action of microorganisms through biochemical process and as the gas is a result of the biological activity it is termed as biogas. Methane gas (CH_4) is the main constituent of biogas (Sahota et al., 2018).

8.2.4.1 Production Process of Biogas

Formation of biogas is an anaerobic process followed by two stages, one is acid formation and another is methane formation. In the very first stage, acid forming bacteria those present in the dung consumes organic compound that present in the waste material and hence acid is formed and termed as acid formation stage. In second stage, methanogenic bacteria act upon organic acid as a result producing methane gas. Main raw materials used in production of biogas are night-soil, poultry litter and agricultural wastes and mainly cattle dung. Advantages of biogas production are highlighted as follows (Divya et al., 2015 Wainaina et al., 2020);

1. One of the eco-friendly methods to produce bio-gas.
2. The raw material required for this process is easily available.
3. The slurry generated after producing biogas is nutrient rich which can be further used for crop production.
4. It keeps environment clean and free from smoky fumes irrespective of other methods of fuel production.

8.2.4.2 Basic Components of Biogas Plants

1. **Mixing Tank-** Mixing tanks are basically designed for collecting feed material and then water is added to form a homogeneous mixture like slurry.
2. **Inlet Pipe-** Inlet pipe works for discharging the substrate.
3. **Digester-** Fermentation process takes place in digester where slurry gets fermented and biogas is formed by the microbial action.
4. **Gas storage dome-** Produced gas is collected in the dome till the time of consumption for further use.
5. **Outlet-** Digested slurry is discharged thorough outlet pipe or any other arrangement made for the same purpose.
6. **Gas connection-** Gas pipe basically used for the purpose of gas utilisation which supplies the gas to lamp or stove (Hubenov et al., 2015).

8.2.4.3 Site requirements for biogas plant installation

Biogas plant installation needs some initial requirements to be made before starting the plant. Following are some of the main points to be considered for constructing the biogas plant.

1. Land should be properly levelled at a higher level than the surrounding background to reduce water spillage.
2. Preferably rough soil should be chosen for construction.
3. For convenient use, site should be nearby cattle shed for easy transportation purpose of raw material.
4. Proper ventilation must be followed for healthy environment.
5. Site should be far away from locality to maintain hygienic conditions.

These are some of the basic points to be strictly followed at the time of constructing any biogas plant.

8.2.5 Brooms from Plastic Waste

Plastic is a versatile material which is used in numerous applications worldwide. There are many new applications of plastic came in front for reducing the increasing peak of plastic waste. Contribution of plastic waste in municipal solid waste have a big share and continuous efforts are been taken for reducing the load of waste plastics. Making brooms out of plastic waste is a unique concept which ultimately serves the purpose of reducing waste generation. Used plastics can be recycled, reused and can be reformed which can be further utilised in the same or in different form according to the type and requirement of the demanded product.

8.2.5.1. Motives

Plastic material is now become an essential part of the human society having diverse applications worldwide (Emadian et al., 2017). Over utilisation of plastics resulted in posing burden on the environment due to mismanaged plastic waste after use and improper disposal practices (Secchi and Zarzur, 1999; Spear et al., 1995; Moharir and Kumar, 2019).

Some important rules to be followed by the citizens for proper management of plastic waste (Moharir and Kumar, 2019). For tackling this waste management issue, from the year January 2018, European Union (EU) has also adopted "European Strategy for Plastics in a Circular Economy" (EU, 2018).

Points discussed under this policy are briefly discussed in this section;

- Plastic waste burden should be reduced to safeguard environment and the citizens.'
- Recycling of plastic waste used in packaging will be done by 2030.
- Single used plastics will be banned or reduced.
- Microplastic use will be restricted.
- Preventive use of plastics will be encouraged.
- Disposing plastic waste in sea will be banned.

These are some mandatory rules to be followed under this policy. Making brooms out of waste plastic is a way of encouraging reuse of waste plastic in a sustainable manner. Such new concepts would be proven helpful in this respect where waste minimization is given more attention to.

8.3 Conclusion and Perspectives

Different ways of tackling an escalating problem of waste management with their advantages are thoroughly discussed in the present chapter. Solid waste management is a topic of global concern and there is a growing need of suggesting better ways to get rid of the present situation for combating the issue of waste management. Traditional methods have many advantages but there is scope of some of the modifications to be done for managing the waste more precisely. There are various methods including landfill, combustion, incineration, pyrolysis, gasification but at the same time all these treatments also pose negative impacts on the environment as well on the human beings. Biological ways are now gained more attention for treating waste material in an environmentally sound manner. Bio-mining being one of them is mostly used and preferred method for extraction of ores by biological means and now-a-day is a very friendly technique in the same respect. Techniques utilising less resources and giving more output is the main target. This chapter finally helps to develop a better understanding of the methods specifically used for treating solid waste.

References

Ahn H.K., Mulbry W., White J.W., Kondrad S.L., 2011. Pile mixing increases greenhouse gas emissions during composting of dairy manure. Bioresour. Technol. 102, 2904-2909. 10.1016/j.biortech.2010.10.142.

Almeida C.M.V.B., Bonilla S.H., Giannetti B.F., Huisingh D., 2013. Cleaner Production initiatives and challenges for a sustainable world: an introduction to this special volume. J. Clean. Prod. 47, 1-10. https://doi.org/10.1016/j.jclepro.2013.03.010.

Azubuike C.C., Chikere C.B., Okpokwasili G.C., 2016. Bioremediation techniques–classification based on site of application: principles, advantages, limitations and prospects. World J. of Microb and Biot. 32 (11), 180. 10.1007/s11274-016-2137-x.

Bharti, V., Vikrant, K., Goswami, M., Tiwari, H., Sonwani, R.K., Lee, J., Tsang, D.C.W., Kim, K.H., Saeed, M., Kumar, S., Rai, B.N., Giri, B.S., Singh, R.S., 2019. Biodegradation of methylene blue dye in a batch and continuous mode using biochar as packing media. Environ. Res. 171, 356–364. https://doi.org/10.1016/j.envres.2019.01.051

Confalonieri R., Jones B, Van Diepen K., Van Orshoven J., 2014. Scientific contribution on combining biophysical criteria underpinning the delineation of agricultural areas affected

Divya D., Gopinath L., Christy P.M., 2015. A review on current aspects and diverse prospects for enhancing biogas production in sustainable means. Renew. Sustain. Rev. 42, 690-699. 10.1016/j.rser.2014.10.055.

Emadian M. S., Onay T. T., Demirel B. Biodegradation of bioplastics in natural environments. J. Waste Manag. 59 (2017), pp. 526-536. DOI: https://doi.org/10.1016/j.wasman.2016.10.006.

EU Strategy on plastics: protecting the environment and citizens from plastic waste, 2018. https://www.interregeurope.eu/policylearning/news/3132/the-eu-strategy-on-plastics-protecting-the environment-and-citizens-from-plastic-waste/

European Commission. Being wise with waste: the EU's approach to waste management, 2010. European Commission. Closing the loop -An EU action plan for the circular economy. COM (2015), pp. 614. Final 2015.

Gaur S., Reed T., Teed T.B., 1998. Thermal data for natural and synthetic fuels. New York, NY: CRC Press; 1998.

Gibbs H.K. and Salmon J.M., 2015. Mapping the world's degraded lands. Appl Geogr. 57, 12-21. https://doi.org/10.1016/j.apgeog.2014.11.024.

Gonzalez M.T., Lopez V.A., Fernandez A.P., Garrido B.R., Cepeda C.T., Mench M., Puschenreiter M., Sabarís C.Q., Garcia F.M., Kidd P.S., 2017. Aided phytostabilisation reduces metal toxicity, improves soil fertility and enhances microbial activity in Cu-rich mine tailings. J Environ Manage. 168 (2), 301-313. https://doi.org/10.1016/j.jenvman.2016.09.019.

He Z., Yang X., Stoffella P., 2005. Trace elements in agroecosystems and impacts on the environment. Elem Medic Biol. 19 (2-3), 125-140. https://doi.org/10.1016/j. jtemb. 2005.02.010.

Hubenov V., Mihaylova S., Simeonov I. Anaerobic co-digestion of waste fruits and vegetables and swine manure in a pilot-scale bioreactor. Bulgarian Chem. Commun. 47 (2015), pp. 788-792.

Huisingh D., Tukker A., Lozano R., Quist J., 2013. Knowledge collaboration and learning for sustainable innovation: an introduction to this special volume (Editorial). J. Clean. Prod. 48, 1-2. https://doi.org/10.1016/j.jclepro.2013.03.050.

Karbowska B., 2016. Presence of thallium in the environment: sources of contaminations, distribution and monitoring methods, Environ Monit Assessm. 188 (11), 640. 10.1007/s10661-016-5647-y.

Kumar, S., Negi, S., Mandpe, A., Singh, R.V., Hussain, A., 2018. Rapid composting techniques in Indian context and utilization of black soldier fly for enhanced decomposition of biodegradable wastes - A comprehensive review. J. Environ. Manage. 227, 189–199. doi:10.1016/j.jenvman.2018.08.096

Lucas N., Bienaime C., Belloy C., Queneudec M., Silvestre F., Nava-Saucedo J. E., 2008. Polymer biodegradation: mechanisms and estimation techniques". Chemosphere. 73 (4), 429–42. 10.1016/j.chemosphere.2008.06.064.

Mandpe, A., Kumar, S., Lakshmikanthan, P., Hettiarachchi, H., 2019a. Mining for recovery as an option for dumpsite rehabilitation: Case study from Nagpur, India. J. Environ. Eng. Sci. 15, 52–60. doi:10.1680/jenes.19.00021

Mandpe, A., Paliya, S., Kumar, S., Kumar, R., 2019b. Fly ash as an additive for enhancing microbial and enzymatic activities in in-vessel composting of organic wastes. Bioresour. Technol. 293, 122047. doi:10.1016/j.biortech.2019.122047

Mandpe, A., Tyagi, L., Paliya, S., Chaudhry, S., Motghare, A., Kumar, S., 2021a. Rapid-in-house Composting of Organic Solid Wastes with Fly Ash Supplementation: Performance Evaluation at Thermophilic Exposures. Bioresour. Technol. 337, 125386. doi:10.1016/j.biortech.2021.125386

Manyapu, V., Mandpe, A., Kumar, S., 2019. Corrigendum to "Synergistic effect of fly ash in in-vessel composting of biomass and kitchen waste" [Bioresour. Technol. 251 (2018) 114–120] (S0960852417321740)(10.1016/j.biortech.2017.12.039). Bioresour. Technol. 277, 225. doi:10.1016/j.biortech.2018.12.031

Moharir R. V. and Kumar S., 2019. Challenges associated with plastic waste disposal and allied microbial routes for its effective degradation: A comprehensive review. J. Clean. Prod. 208, 65-76. https://doi.org/10.1016/j.jclepro.2018.10.05.

Moharir R. V., Gautam P., Kumar S., 2019. Waste Treatment Processes/Technologies for Energy Recovery, Chapter 4. Current Develp. in Biot. and Bioe., (2017), pp. 55-77. https://doi.org/10.1016/B978-0-444-64083-3.00004-X.

Moharir, R. V., Kumar, S., 2019. Challenges associated with plastic waste disposal and allied microbial routes for its effective degradation: A comprehensive review. J. Clean. Prod. 208, 65–76. https://doi.org/10.1016/j.jclepro.2018.10.059

Nagendran R., Selvam A., Joseph K., Chiemchaisri C., 2006. Phytoremediation and rehabilitation of municipal solid waste landfills and dumpsites: a brief review. Waste Manag. 26 (12), 1357-1369. 10.1016/j.wasman.2006.05.003.

Otte M.L and Jacob D.L., 2008. Mine area remediation. Encycl. Ecol. 2397-2402. 10.1016/B978-008045405-4.00073-2.

Perez-Lopez R., Delgado J., Nieto J.M., Marquez-Garcia B., 2010. Rare earth element geochemistry of sulphide weathering in the São Domingos mine area (Iberian Pyrite Belt): a proxy for fluid–rock interaction and ancient mining pollution. Chem. Geol. 276 (1-2), 29-40. https://doi.org/10.1016/j.chemgeo.2010.05.018.

Rao, M. N., Sultana, R., & Kota, S. H., 2017. Municipal Solid Waste. Solid and Hazardous Waste Management. pp. 3-120. 10.1016/b978-0-12-809734-2.00002-x.

Robert R., 1992. On-Farm Composting Handbook, Plant and Life Sciences Publishing, Cooperative Extension, Ed. (June 1992),

Rodriguez L., Azcarate J.A., Rincon J., 2009. Heavy metal distribution and chemical speciation in tailings and soils around a Pb–Zn mine in Spain. J Environ Manage. 90 (2), 1106–1116. https://doi.org/10.1016/j.jenvman.2008.04.007.

Sahota S., Shah G., Ghosh P., Kapoor R., Sengupta S., Singh P., 2018. Review of trends in biogas upgradation technologies and future perspectives. Bioresour. Technol. 1, 79-88. https://doi.org/10.1016/j.biteb.2018.01.002.

Secchi E. R. and Zarzur S., 1999. Plastic debris ingested by a Blainville's beaked whale, Mesoplodon densirostris, washed ashore in Brazil. Aquat. Mamm. 25(1), 21-24.

Silvestre B.S. and Silva Neto R., 2014. Are cleaner production innovations the solution for small mining operations in poor regions? The case of Padua in Brazil. J. Clean. Prod. 84, 809-817. https://doi.org/10.1016/j.jclepro.2014.01.097.

Silvestre B.S., 2015a. Sustainable supply chain management in emerging economies: environmental turbulence, institutional voids and sustainability trajectories. Int. J. Prod. Econ. 167, 156-169. https://doi.org/10.1016/j.ijpe.2015.05.025.

Silvestre B.S., 2015b. A hard nut to crack! Implementing supply chain sustainability in an emerging economy. J. Clean. Prod. 96, 171-181. https://doi.org/10.1016/j. jclepro.2014.01.009.

Spear L.B., Ainley D.G., Ribic C.A., 1995. Incidence of plastic in seabirds from the tropical pacific, 1984-1991: relation with distribution of species, sex, age, season, year and body weight. Mar. Environ. Res. 40 (2), 123-146. https://doi.org/10.1016/0141-1136(94) 00140-K.

Wainaina, S., Awasthi, M.K., Sarsaiya, S., Chen, H., Singh, E., Kumar, A., Ravindran, B., Awasthi, S.K., Liu, T., Duan, Y., Kumar, S., Zhang, Z., Taherzadeh, M.J., 2020. Resource recovery and circular economy from organic solid waste using aerobic and anaerobic digestion technologies. Bioresour. Technol. 301, 122778. https://doi.org/10.1016/j. biortech.2020.122778 by specific constraints: Methodology and factsheets for plausible criteria combinations. InJRC Science and Policy Reports, EUR-26940 EN, Terres JM, Hagyo A, Wania A, editors, 2014, Publications Office of the European Union, Luxembourg, pp. 81.

Yasar D., 2018. Comprehensive Energy Systems. Chapter 1.22, Biofuels, 1(2018). dx.doi. org/10.1016/B978-0-12-809597-3.00125-5.

Yavitt J.B., 2015. Land use history determines non-native earthworm impacts on atmospheric methane consumption in forest soils, central New York State. Can. J. of Soil Sci. 95, 321-330. https://doi.org/10.4141/cjss-2014-073.

Zhao X.G., Jiang G.W., Li A., Wang L., 2016. Economic analysis of waste-to-energy industry in China. Waste Manage. 48 , 604-618. https://doi.org/10.1016/j.wasman.2015.10.014.

9

Environmental Impacts of Waste Management

Anoop Jaiswal[1,3], Sneha Singh[3], Ashootosh Mandpe[1,2] and Sunil Kumar[1]

[1]CSIR National Environmental Engineering Research Institute (CSIR-NEERI), Nehru Marg, Nagpur 440 020
[2]Department of Civil Engineering, Indian Institute of Technology Indore Indore 453 552, India
[3]Indian Institute of Forest Management, Bhopal-462 003.

ABSTRACT

Waste management has always been a burning environmental challenge. In the present scenario, the accelerated waste production and its characterization complexity has added to the country's trouble. The worst affected are the developing countries that struggle to manage their waste due to overpopulation, land scarcity for disposal, or technological advancement. Waste, in general, can be broadly categorized as Hazardous and Municipal solid Waste. This broad categorization accounts for the major proportion of total waste production. In general, Solid waste is non-toxic. However, at times it is found to contain a mixture of inert and toxic components, which makes the safe handling, timely management and quick disposal are an absolute necessity to avoid any miss-happening at a great scale. Solid waste management approaches differ in each country due to cultural, technological, and economic constraints. Broadly these approaches are open dumping, landfills, incineration. However, each of these approaches has a different environmental impact, and no approach is completely environment friendly as these contribute to environmental pollution via emissions in one way or another. The impression of this, at times, are visible on human health and floral productivity. The presented chapter is a take on the different solid waste management approaches and their impact on the environment.

Keywords: Waste management, Hazardous waste, Solid waste, Open Dumping, Landfills, Incineration.

Introduction

With ever-increasing population growth, waste management has become a major challenge threatening environmental sustainability. The total solid waste generation was recorded to be 1.3 billion tonnes per year by the world's cities in 2012, with an ecological footprint of 1.2 kilograms per person/day. With the intensity with which the population growth is happening coupled with urbanization, predictions are made that by the end of the year 2025, 2.2 billion tonnes of municipal waste will be generated (The World Bank, 2018).

Studies have shown that one of the major reasons for this waste menace is the non-segregation of waste before disposing. The inability to ensure the segregation of waste becomes a hindrance to the safe disposal of waste, which results in the release of toxic pollutants into the environment. In developing countries, it has also been observed that allocated sites for landfilling are often used for open dumping of waste without, resource recovery, which generates leachate and releases gases in the natural boundaries (Mandpe et al., 2019a; Dhar et al., 2017).

Untreated waste (a mixture of non-biodegradable and biodegradable waste) stays in the dumpsites for months, which further adds to the issue. The solid waste menace is not only because of the piling up of the old waste but also due to the non-effective handling of present waste. The lack of infrastructure and SWM delivery mechanisms in developing countries can be dealt with by proper planning and a well-coordinated framework of the city development plan.

Based on several factors, whether land availability, per capita generation, technological advancement, or economic feasibility, every country opts for a waste disposal method. As not a single method can be used to dispose of all the waste, a set of methods is used. Generally, these methods are open waste dumps, landfills, and incineration. However, these approaches do have associated environmental implications. In developing countries, the selection of improper waste dumping practices contributes to serious health implications, threatens the safety of the ecosystem, and result in causalities via the outspread of diseases. At times, a waste disposal site serves as the vector of the disease.

To have sustainable cities, waste management is a necessity. It has also become a goal for most of developing countries (The World Bank, 2018). The solid waste management scheme is a set of complex systems of economic, environmental, social, political, legal, operational, and technical nature. Rational decisions cannot be taken and applied unless a complete and accurate picture of the existing problem is available and a hierarchical classification of the system components can be produced (Hadjibiros et al., 2011). To identify the existing solutions, it is important to understand the process as a whole, and the process is discussed in the sections ahead.

9.1 Air Pollution Associated with Different Waste Management Processes

Increased population growth rate, urbanization, economic expansion, and industrialization has resulted in increased human consumption of products and services, which in turn has led to increased accumulation and generation of waste (Isaifan & Alyasi, 2018). Affluence and shift in technology from returnable packaging to non-returnable packaging are also found as major factors aggravating the tendency of consumers to discard. If this waste remains untreated and unmanaged, it can cause various environmental impacts and health hazards (SN & Chadar, 2017). The processes involved in the management of this waste generate a significant fraction of global air pollution. In one of its reports, IPCC declared post-consumer waste as a small but significant contributor to greenhouse gas emissions. Methane from the landfill and wastewater and N_2O are reported to be the major emissions from the waste sector contributing to greenhouse gas emissions (Ahmed et al., 2007). Thus, making solid waste management an important and notable source adding to the existing problems of air pollution. There are many solid waste management techniques, popular being landfill, incineration, composting, waste collection, transfer stations, and recovery (Weitz, Barlaz, Ranjithan, Brill, Thorneloe, & Ham, 1999). Each of these systems has its carbon flows, making them accountable for emitting different amounts and varieties of greenhouse gases. CH_4 emissions from Landfills and CO_2 emissions from the fossil carbon component of incinerated waste are considered the most popular gaseous C emissions from solid waste.

Nevertheless, few other solid waste sources contribute to GHG. These are CH_4 from anaerobic conditions within composting operations, CH4 leaks from anaerobic digestion reactors, N_2O from nitrification in compost piles, and volatile organic compounds from compost activities, metals from open burning and emissions from collection and transportation of waste to transfer & treatment sites (Mohareb, MacLean, & Kennedy, 2011). Emissions of these harmful organic and inorganic gaseous pollutants cannot be avoided completely even after following the scientific waste management techniques (Majumdar & Ganguly, 2018). Thus, responsible collection, management, and disposal of solid waste have arisen as a major challenge for municipalities, local bodies, and the nation as a whole.

There are many uncertainties associated with direct and indirect emissions and mitigation potentials for the waste sector. Proper collection, analysis, and validation of data related to the waste sector are missing. Waste management decisions often made locally are influenced by various factors, including local economics, national policies, public perceptions, regulatory restrictions, and infrastructural requirements (Kumar et al., 2018). Such factors, along with no proper concurrent quantification of GHG emissions and GHG mitigation, deeply undermine the capacity of the waste sector in contributing to the minimization of greenhouse gas emissions in the environment (Ahmed et al., 2007). Thus, consistent national definition of waste across the nations, synchronized local and international data collection, planned maintenance of national GHG inventories, regularized data analysis, and field validation of models could play an instrumental role in planning effective mitigation of GHGs and other gaseous pollutants from this sector. Thus, to promote public health, quality of life, conservation of natural resources, and availability of benefits of renewable energy, establishing an affordable, effective and sustainable waste management system is a must.

9.3.1 Solid Waste Transportation Associated Emissions

Emissions are not released just from the waste disposal sites carrying disposal processes but also from the transportation or movement of these wastes. Many studies assert that waste transportation plays an important role in adding pollutants, especially greenhouse gases, to the environment. Diesel fueled vehicles used for waste transportation have been shown to discharge pollutants like carbon dioxide (CO2), carbon monoxide (CO), hydrocarbon (HC), oxides of nitrogen (NOx) and particulate matter (PM). Transportation of waste can carry significant global warming potential compared to global warming potential caused by entire waste management system (Sharma & Chandel, 2018). Out of these, carbon dioxide (CO2) and oxides of nitrogen are potent greenhouse gases giving rise to the problem of global warming.

Other important local non-greenhouse gas pollutants released during the waste collection and transportation processes are oxides of sulfur (SOx) (Weitz, Thorneloe, Nishtala, Yarkosky, & Zannes, 2002). Sea transportation of waste results in a worse situation causing environmental damage through acidification. The emissions of sulfur oxides and nitrogen oxides during the process from the barges are accountable for such consequences (Tan & Khoo, 2012). The emissions from this category of disposal mechanism have increased due to the increase in waste being generated (Weitz, Thorneloe, Nishtala, Yarkosky, & Zannes, 2002).

9.3.2 Emissions from Dump Sites

The dumpsites are almost identical to sanitary landfills as far as the nature of emissions from the waste management system is concerned. Still, the major difference is brought by the regulatory status shared by both of them (Heimlich). While the latter operates as a well-engineered, controlled, protected and monitored waste management technique, mostly regulated by municipalities or local government bodies, the former represents an unplanned, unregulated land disposal site which is generally not checked for its safety and environmental protection (Federal Facilities Environmental Stewardship & Compliance Assistance Center, 2017). Their unknown site capacity mainly characterizes these, indiscriminate disposal of waste, little or no site preparation, no leachate management, no gas management, no compaction of waste, and no record keeping of activities (Mavropoulos, 2013). These dumpsites are majorly known to produce methane, bad odour, dust, and microbial air pollution (Isaifan & Alyasi, 2018). Dust associated with dumpsite activities comprises fine and coarse particles, falling in the PM2.5 and PM10 categories, respectively. This particulate air pollution is sensitive to children, the elderly and people with pre-existing lung and heart disease. The incomplete combustion of dumpsites bio components releases Black Carbon, particles of diameter less than 2.5μm, known for settling in the deeper parts of the lungs. The same has been proved to cause short term changes in health (Mavropoulos, 2013).

Apart from this, the open burning of such dumpsites, practised popularly to reduce the waste volume, bringsmany environmental and public health effects. It releases uncontrolled smoke into the environment, which eventually destroys the ozone layer by causing air pollution. Unlike landfills, dumpsites indiscriminately accept all kinds of unchecked hazardous and non-hazardous wastes. These can be Municipal solid waste, E-waste, or Health care waste. The unmonitored garbage disposed of in these dumpsites, on being burnt, can release air toxins like mercury, nitrous oxide (at high temperatures), furans, and dioxins into the atmosphere (Conserve Energy Future, 2020). Metals like barium, antimony, beryllium, cadmium, arsenic, chromium, lead, mercury, manganese, phosphorus, and titanium are also liberated during the process. Dioxin, a carcinogen associated with congenital disabilities, is one of the most harmful pollutants released during the open burning of dumpsites. Municipal Solid Waste component of dumpsites, on being burnt, can release hexachlorobenzene (HCB), a highly persistent air toxin and human carcinogen, whose low-level, long-term exposures can damage a developing fetus, a functioning liver or kidney and can result in skin irritation and fatigue. Burning of pressed wood products, paints, coatings, and fibreglass insulation discharges formaldehyde in the atmosphere, short term exposure of which causes watery eyes, burning sensation in eyes and throat, nausea, coughing, chest tightness and wheezing.

Moreover, its long-term exposure can lead to cancer. Other air pollutants like hydrogen chloride gas, carbon monoxide, and Black Carbon are also released through the burning of plastic related waste, incomplete burning of waste, and open burning of waste, respectively (Mavropoulos, 2013). Various hazards resulting from these dumpsites are mainly catalyzed by the improper designing decisions associated with them. These mainly include the absence of leachate collection and treatment mechanism, shortage of cover and liners, faulty site selection, and poor site design (Yadav, Kumar, & Singh, 2018).

9.3.3 Emissions from Aerobic Composting

Composting is an aerobic waste disposal method that reduces the emission of methane during the breakdown or degradation of waste products. Anaerobic decomposition of waste, happening in the absence of oxygen and the presence of microbes, releases methane into the atmosphere. However, these methane-producing bacteria are not active in the presence of oxygen. Thus, aerobic composting practice, in comparison to anaerobic fermentation, is a more efficient and environmentally superior method to reduce the greenhouse gas emissions arising out of solid waste (Department of Primary Industries and Regional Development's Agriculture and Food division, 2018). Although mixed results were found in some instances related to greenhouse emissions made from composting operations and its counterpart anaerobic degradation mechanisms, in general terms, the greenhouse emissions associated with the latter tend to be higher than those associated with the former (Lou & Nair, 2009). Composting breaks down the waste aerobically into carbon dioxide, humic fraction, and water. The residual compost also contains some carbon. Apart from reducing methane, there are many advantages known to this method of solid waste disposal. The compost produced can have many potential applications in agriculture, horticulture, soil stabilization, and soil improvement (Ahmed et al., 2007). It increases the soil buffering capacity, soil moisture-holding capacity, and capacity to retain soil fertilizer (Paliya et al., 2019). The addition of an organic matter source to the soil increases the pool of nutrients, improves the soil structure, and stimulates the biological activity in it. Thus, composting leads to direct carbon sequestration through the compost and indirect carbon sequestration through increased biomass of plant root systems (Department of Primary Industries and Regional Development's Agriculture and Food division, 2018).

In addition to CO2, several other emissions are also discharged during the process of composting. The poorly managed composting process can lead to the formation of semi-aerobic and anaerobic conditions, which in turn can produce greenhouse emissions like N2O and CH4, respectively (Mandpe et al., 2020a). Even well-managed systems have been indicated, in certain cases, to release CH4 and N2O. (Ahmed et al., 2007). Recent studies have shown that composting high organic content wastes result in the production of N2O (He, Inamori, Mizuochi, Kong, Iwami, & Sun, 2001). Volatile compounds like carbonyl compounds, alcohols, esters, terpenes, sulfur compounds, and others have been reported to be found in aerobic composting waste gases. Acetone, 2-propanol, ethanol, ethyl acetate, and limonene are some of the predominant volatile organic compounds released during the aerobic composting process of biowaste (Smet, Langenhove, & Bo, 1999). These volatile organic compounds are potentially odorous and cause aerobic composting to generate odour, thereby making it a community concern (Eitzer, 1995). Such malodor problems present an important environmental issue in countries with a high population density. Volatile sulfur compounds (VSCs) are identified as the major odorants emitted by waste composting treatments. These are volatile organic compounds comprising sulfur and reduced sulfur compounds (RSCs), characterized by low detection threshold and strong odour properties, strongly capable of adding to odour pollution at lower concentrations. The predominant VSCs detected in composting treatments of municipal solid waste and kitchen waste are dimethyl sulfide (DMS), methyl mercaptan (MM), hydrogen sulfide (H2S), carbon bisulphide (CS2) and dimethyl disulfide (DMDS) (Zhang, Schuchardt, Li, Yang, & Yang, 2013).

9.3.4 Emissions from Landfill Sites

Landfills are among the most important regulated and monitored solid waste management techniques, especially in developing nations. They are designed to isolate the trash from the surrounding environment, mainly air, groundwater, soil, rain, and residential areas (Advanced Disposal, 2020). This isolation is attained through operating practices like compacting, bottom lining, and a thick covering of waste with soil (Environmental Protection Agency, 2020). This compacting and covering of waste eliminates the oxygen and eventually breaks it down in an anaerobic process (Environment Victoria, 2013). They are anaerobic digesters that can continue to store GHG emissions even several decades after the waste disposal, thus making them inefficient. Solid waste dumped in landfills initially undergoes aerobic decomposition, which during later stages get replaced by anaerobic conditions marked by strong activity of methane-producing bacteria over waste products (Landfill Methane Outreach Program (LMOP)). After starting off aerobic, landfills fairly soon become anaerobic in large parts due to the placement of more and more on top of and around the earlier waste. This body of waste ceases the ability of oxygen to diffuse into the mass of waste, thus making the decomposition process anaerobic (Waste Decomposition and Degradation of Waste in Landfills, 2019). Various gaseous compounds are produced in landfills due to the degradation of matter present in the waste. The same is stated to affect the adjoining regions by posing air pollution, health threats, and landscape degradation. These mainly comprise methane, carbon dioxide, hydrogen, hydrocarbons, dust, odours, and microorganisms. Methane (CH4) is the strongest GHG emitted by landfills and has a Global Warming Potential of 28 to 36 times more than CO_2 over 100 years (Environmental Protection Agency, 2017). Minor emissions of carbon dioxide are also produced by this waste management system (Ahmed et al., 2007).

Quantitatively, methane and carbon dioxide are the most important emissions from landfills. Ammonia, nitrogen, hydrogen sulfide, monocarboxylic acids, sulfur, and many volatile fatty acids are a few of the malodorous compounds formed in landfills due to various biological and chemical processes being carried on in the waste. They are responsible for the irritating bad odour generated in the neighbouring atmosphere. Landfills can also release environmentally toxic compounds like mercury, silicon compounds, carbon monoxide, and aromatic compounds like benzene, toluene, xylene, ethylbenzene, and propyl benzene. Inhalation of even a small amount of mercury can harm the kidneys and respiratory system (Bausback, 2016). Active landfills have also been asserted as the highest emitters of bioaerosols, compounds containing different microorganisms that can trouble the population by inducing respiratory disorders, digestive disorders and skin diseases (Isaifan & Alyasi, 2018). The gaseous compounds emitted from landfills have different impacts and different life spans and act on different spatial and time scales. While the dust, aerosols, aromatic hydrocarbons, hydrogen sulfide, carbon monoxide, mercury, and Si compounds originating from landfills create unwanted disturbances locally, gases like methane, carbon dioxide, ammonia aggravate the problem of greenhouse gases globally. On the local scale, the effect is mainly posed on vegetation, odour, and health. Whereas, globally, impacts posed chiefly comprise of global warming and environmental persistence. Dust and odour can be a transient phenomenon, but methane, on the other hand, is a short-term acute hazard having far-reaching and long term effects (Fischer, Maurice, & Lagerkvist, 1999).

9.3.5 Emissions from Incineration

Incineration of solid waste is one of the solid waste management techniques which is increasingly gaining application in the developed nations, owing to the introduction of stricter landfill directives. It is the process of burning solid waste in a properly designed furnace under appropriate temperature and operating conditions (SN & Chadar, 2017). This technique avoids most of the GHG generation in comparison to the Landfill technique, which produces uncontrolled emissions of CH_4. CO_2 is the chief emission released from the fossil C component of waste being subjected to incineration. In contrast, the CO_2 emissions from the biomass/biogenic proportion of the waste are not accounted for GHG inventory, as prescribed by IPCC. Unlike developed countries like Japan, Denmark, Luxembourg, France, Sweden, the Netherlands, and Switzerland, who have high rates of incineration, developing nations are affirmed to show controlled rate and frequency of incineration. The technique's high operating and capital costs, along with an unfavourable history of failed projects, are reported to be the major reasons behind the limited application of this technique (Ahmed, et al., 2007).

The incineration mechanism results in the release of climate-relevant emissions. Carbon dioxide (CO_2), Nitrous oxide (N_2O), Ammonia (NH_3), oxides of nitrogen (NO_x), and organic carbon (C) are mainly emitted during incineration of solid waste. Methane (CH_4) is not released during normal operations of incineration. However, a smaller fraction is emitted only in the rare cases of the waste remaining in the waste bunker. The incineration of 1 Mg of heterogeneous municipal waste in MSW incinerators discharges about 0.7 to 1.2 Mg of carbon dioxide (CO2). Out of this, 0.33 Mg of CO_2 belongs to the fossil carbon category, produced from the combustion of non-biogenic waste components like plastics and synthetic textiles. With regards to Nitrous oxide (N_2O), emission levels of 1 to 12 mg/m3, with an average of 1 - 2 mg/m3 have been specified to be released from an incineration activity (Johnke). Though the CO_2 released through the incineration process enhances the greenhouse effect, no attempts are made to minimize such emissions. The reason for this can be attributed to the relatively small contribution made by this source to the total CO_2 emissions.

Other releases from this process include sulfur dioxide and oxides of nitrogen, both being important sources responsible for developing urban smog and causing acid precipitation. Incineration of solid waste products like batteries, thermometers, and electrical switches, containing significant mercury concentrations, leads to vaporization of the metal during the process, which can have dangerous implications. Most belligerent emissions from this waste disposal mechanism are released when chlorine-containing organic materials and chlorinated plastics are combusted. Combustion of PVC (Polyvinyl chloride) and PCBs (Polychlorinated biphenyls) liberates highly toxic dioxins and furans (science.jrank, 2020). Incineration of more complex wastes like hospital waste is marked by stack emissions of particles, carbon monoxide, hydrochloric acid, ethylene, propane, ethane, and propylene (Allen, Brenniman, & Darling, 2012). Studies convey that incineration, when used for disposal of electronics industry waste like personal computer motherboards, keyboards, and cases, release significant metal emissions of copper lead and antimony (Stewart & Lemieux, 2003)

Considerations for Future

Many policies, measures, and practices, which have been demonstrated for their sustainability, are being followed worldwide to improve the existing solid waste management processes. Policies like Unit Pricing (PAYT/Pay As You Throw) have been widely adopted to reduce the load on landfills and increase the tendency of waste recycling. Other measures being practised in different parts of the world including local subsidies for collection of recyclables, green procurement or purchase of recyclable products and domestic composting of biodegradable wastes (Negi et al., 2020). Waste, if managed properly, can become a resource. Waste minimization, recycling, reuse, and energy recovery have been cited as the sectors of solid waste management where long term and more profound changes should be brought. Though incineration has been established as an efficient technique that can capture and use the energy value of post-consumer waste, it is not practised widely in developing countries. The high capital and operational investments associated with the technique refrain its sustainable use in these nations. Kyoto mechanisms like Joint Implementation and Clean Development Mechanism (CDM) are trying to bridge this gap by providing strong economic incentives for improved landfilling practices and landfill CH4 recovery. Open dumping and open burning practices of waste disposal are positively being replaced by landfill gas recovery projects in the developing and EIT countries. Utilization of this landfill CH_4 can offer an economic mitigation solution. To increase the adoption of more efficient methods other than landfill disposal, many countries are implementing subsidies for the construction of incinerators, tax exemptions on electricity generated by incineration of waste, tax exemptions on waste disposal with energy recovery, and introduction of landfill taxes. Mechanical Biological Treatment (MBT) is now being used in Germany, Italy, and Australia. This process subjects mixed waste to a series of mechanical processes (sorting, shredding, and cutting) to produce waste fraction, later recycled or biologically treated through anaerobic digestion and/or aerobic composting. In comparison to landfilling, MBT theoretically can reduce generation of CH_4 by 90% and increase the rapid waste stabilization process. Although anaerobic digestion technologies incur low capital costs compared to incineration, their potential to mitigate GHG is limited as compared to incineration. Their potential to mitigate GHG is limited compared to incineration and landfill CH_4 recovery. Aerobic composting of source-separated biodegradable streams is being successfully deployed in many countries. Anaerobic digestion can be more advantageous to aerobic composting in terms of energy benefits (biogas), reduced process time, and reduced volume of residuals but becomes disadvantageous on cost parameters. Composting can provide a more affordable, sustainable, and environmentally acceptable solution to controlled landfilling in developing countries were more labor-intensive simple technologies are preferred and used for selected biodegradable waste disposal (Ahmed, et al., 2007). Thus, the future of solid waste management can see more effective, innovative, and incremental changes in the current waste disposal methods.

9.4 Water Pollution allied with Varied Solid Waste Clearance Practices

Water is the necessity of every living being to survive and flourish. Of the total water available on the planet, only 3% accounts for the freshwater consumed by humans. The changes driven by industrialization, urbanization, and massive population growth have caused some long-lasting changes in the surroundings.

These changes have taken a toll on human life as well via deteriorating the physical boundaries and outspread of disease via contamination of natural resources that humanity depends on. Water is one such resource impacted by the humane way of disposing of the waste. The importance of waste dumping was understood when the impacts become visible on the environment. Groundwater quality emerged as an important environmental issue confronting the world's population in the last two decades (Moench, 1996).

Several methods have been adopted to evolve strategies for better waste management (Kumar and Gaikwad, 2004). Solid waste clearance practices such as incineration and landfills and the conventional practice of open site dumping, which are in practice in the existing scenario, are not free from pollution.

These practices do have specific impacts on the environment.

The selection of waste disposal mechanism by overlooking the waste characteristics is another issue for the groundwater and surface water pollution. The groundwater quality problem has become significant nowadays (Kumar and Sinha, 2010). In the developing countries, unprecedented population growth has caused many cities struggle in infrastructural services viz., solid waste management, sewerage, and water supply. Be it open waste dumping near water bodies that contaminate the surface water, or the groundwater pollution driven by the seeping leachate from the landfills; both result from not so stringent waste management policies and frameworks. Several other factors, such as economic constraints, land availability, etc. also play an important role in decision making.

9.4.1 Contamination of water at Open Dump Sites

Water Contamination has emerged as a growing challenge across the globe, which is posing a serious threat to human and environmental health as a whole (Ocheri *et al.,* 2008). Worst affected are the developing countries where this challenge has become apparent because of the anthropogenic activities, which comprise rapid population growth, industrialization, and other productive activities (Uting *et al.,* 2007). Studies have shown that there exists an association between stream water quality and human activities, particularly since the proclamation of the Worlds Drinking Water Supply Decade in the 1990s (Ocheri *et al.,* 2008). Open dumpsites are one such structure created by humans for disposing of solid waste, which encourages open dumping of wastes. Open dumping of waste is an age-old practice the oldest and commonest method of getting rid of solid wastes practised in many regions across the globe (Nzeadibe, 2009). Open dumpsites are a source of solid waste led pollution responsible for threatening the quality of surface water. With the tremendous increase in population, the world has experienced a phenomenal increase in the quantity of solid waste generation and its disposal in open dumpsites (Aliyu, 2008). Worst affected are the areas where open dumpsites are located close to the surface water bodies.

There is a strong correlation between the degree of stream water pollution, and the frequency of waste dumps near the water body (Aliyu, 2008). The waste in dumpsite changes by interacting with the wastes of different natures, moisture, and the atmosphere, which further makes it a potent pollutant. There are several ways by which the pollutants from the open dumping sites can reach the water bodies depending on the proximity between the two. Pollutants may reach the surface water via wind or can be washed away by the rains, or by an accidental introduction by a living being in the water body.

In addition to the contamination of surface water, open dumpsites even account for groundwater pollution. Open dumping sites undergo a biological and chemical change led by the interaction with the moisture in waste and the atmosphere. This interaction results in the weathering process (Sabahi et al., 2009). The hydrologically mediated changes in the open dumping site pave the way to the formation of leachate. Leachates serves as a source of heavy metal pollution to both soil and aquatic environments (Ali & Abdel-Satar, 2005). Leachate also poses detrimental effects on human health via soil, crop, and water contamination. (Bahnasawy et al., 2011). Leachate gets readily absorbed by the soil, and it moves to the groundwater & the aquifer system under the action of gravitational pull. Upon reaching groundwater, it fiddles with the Physico-chemical nature of water.

Studies have reported an increased risk of illness to those exposed to pollutants migrating from waste dumps. It is even believed that pollution level and water associated infections are proportional to the duration of exposure to contaminants (Uting *et al.,* 2008). Leachates from dumpsites constitute a source of heavy metal pollution to both soil and aquatic environments (Ali & Abdel-Satar, 2005). This may have detrimental effects on soils, crops and human health (Bahnasawy et al., 2011).

9.4.2 Contamination of water at Landfills

Landfilling is considered to be one of the oldest operational methods of municipal solid waste disposal, which is still preferred due to its effectiveness and economics. However, high production of waste and scarcity of land for landfills is an issue in most developing countries with high population density (Sadek and El-Fadel, 2000).

The solid waste dumped in landfills is commercial, industrial, municipal, and agricultural waste. Landfills are generally lined from inside to prevent soil and groundwater contamination. However, in the case of poorly designed landfills, these lines are either not there or are broken, thus serving as the source of groundwater, soil, and air pollution (Jhamnani and Singh, 2009). Leachate is formed when the moisture content in the landfill is sufficiently high to initiate the liquid flow. Interaction of solid waste with moisture content generates the leachate in the landfill, and the contaminants from the solid waste are carried by the percolation of the leachate. Contaminants associated with the waste can be present in the form of byproducts of biological and/or chemical reactions or can be in the form of dissolved or suspended materials associated with the disposed of waste in the landfill. The strength of the leachate is the function of chemical/biological processes that contaminants in the leachate go through.

The rate of leachate production and its characteristics are the function of many factors such as Site hydrology, age of landfill, moisture and temperature conditions, oxygen availability, the composition of solid waste, particle size, and degree of compaction. Based on the constituents, leachate consists of non-conservative and conservative constituents. Non-conservative constituents are primarily organic, whereas conservative constituents include heavy metals, sulfide, chloride, etc. Metals are often precipitated within the landfill and are mostly found in the leachate in high quantity, except for iron. Non-conservative constituents tend to decompose and stabilize with time during the stabilization of landfilled wastes. Conservative constituents, on the other hand, remains even after the waste stabilization. (Kamboj et al., 2013)

The use of leachate contaminated groundwater is the most commonly reported danger to human life (Rajkumar et al., 2010). Therefore, to get rid of the ground water pollution and prevent the groundwater resources from damage, urgent steps for detailed identification and abatement are necessary (Jain et al., 2004).

9.4.3 Contamination of water by Incinerated wastes

Incineration is a waste management approach developed to answer the challenge posed by ever-increasing waste generation. This waste management approach is extensively applied to municipal solid waste, medical waste and hazardous waste. However, the combustion reaction taking place in the Incinerators forms ash along with carbon and water vapours. In some instances, when combustion doesn't reach its maximum limit, harmful substances can be produced. The process type, nature of waste, and combustion conditions determine the concentration and characteristics of the contaminants in the waste stream (Waste Incineration and Public Health, 2000)

The process emissions from the incinerator can be categorized as **Particulate matter/fly ash**

Particulate matter is primarily the non-combustible content in the flue gas. This non-combustible part contains inorganic ash content in the waste, and carbonaceous soot produces by the combustion process in the incinerator. The inorganic-ash fraction of the particulate matter is formed by the mineral matter and metallic species, which are left untouched during combustion and exit the chamber as fly ash. On the other hand, soot is the unburned carbon part left by incomplete combustion. Upon release in the atmosphere, the particulate matter fiddles with the physical characteristics of the surface water, and depending on the nature of waste, it can also make water unfit for use.

Acid gases

Acid gases are part of particulate emissions. These are potential environmental polluters as they can form acids upon interaction with water vapor. The acid thus formed gets condensed in the atmosphere and returns to the surface via precipitation (acid rain), or it can simply dissolve in water making the water acidic. Acid gases comprise the oxides of nitrogen, oxides of sulfur, hydrogen chloride, hydrogen bromide, hydrogen iodide, and hydrogen fluoride. These gases are released into the environment via uncontrolled emissions ranging from a hundred to thousand parts-per-million-by-volume. Production of free halogens such as bromine, chlorine, and iodine are also observed at low concentrations from the combustion of wastes containing the compounds of those elements.

Heavy metals

Heavy metals in the waste are conserved from the incineration process. It has been observed that metallic elements having high vapor pressures, or the metals capable of forming the compounds with high vapor pressures, can be converted directly to the vapour phase in the combustion chambers. Upon cooling, these metals condense. In some volatile metal compounds, the same phenomenon is observed. For example, Mercury and lead are converted to the vapor phase during the combustion process. Upon interaction with the water vapor in the atmosphere or the water body on the earth's surface, the exhaust containing heavy metals may form compounds of lead and mercury. Upon consumption, the contaminated water leads to bioaccumulation in the floral and faunal species.

Products of Incomplete Combustion

PICs are free radicals formed by the breakdown of organic compounds in the combustion unit. These are not combusted and released via the flue gas. The PICs combine to form many organic compounds. PIC emissions are the function of combustion conditions. Depending on the temperature, some of the heavy organic constituents can condense onto fine particles. PICs include simple compounds (e.g., methane, ethane, acetylene, and benzene), dioxins and furans, partially oxidized organic compounds (e.g., acids and aldehydes), and polycyclic aromatic hydrocarbons.

The process emissions by the incineration seem to be potent enough to fiddle with the physical, chemical, and biological characteristics of surface water. It makes the use of incinerators coupled with advanced methodologies to keep the emission led to contamination under control.

9.5 Soil Pollution due to Solid Waste Diverse Disposal Methods

Disposal of solid waste is an extensively throbbing problem in both urban and rural areas of developed and developing countries (Abdel-Shafy & Mansour, 2018). The resident's progress and inadequate solid waste disposal and management possibilities harvest huge amounts of trashes in abundant forms, which influences the environment. The potential risk posed to the soil by this waste is considered as the biggest environmental problem associated with its disposal technique and disposal site. (Taddese, 2019).

Different disposal mechanisms of solid waste exert varying negative consequences on the soil at disposal sites and in its vicinity areas (Ali, Pervaiz, Afzal, Hamid, & Yasmin, 2014). Leachate generation is an inevitable consequence of practising solid waste disposal through the landfill method. Leachate at the disposal site and its migration away from the landfill boundaries cause grave environmental concerns of soil pollution, vegetation damage, landfill settlement, unpleasant odour, and groundwater pollution. Leachate significantly affects the chemical and geotechnical properties of soil (Pillai, Peter, M, & S., 2014). The soil at disposal sites has been studied and analyzed to exhibit high pH, Total Dissolved Solids, and Electrical Conductivity in comparison to its control units (Ali, Pervaiz, Afzal, Hamid, & Yasmin, 2014). It can modify the soil properties and notably alter the behaviour of soil, thus affirming that soil waste dumps in landfills affect the engineering and chemical properties of soil. It reduces soil strength and eventually undermines its usefulness as a foundation material (Pillai, Peter, M, & S., 2014).

Open dumping of waste comprising of indiscriminate waste disposal from human settlements, small industries, commercial units, hospitals, and clinics, is a very common and low budget solution to solid waste management in developing countries. Such practice releases heavy metal contaminants like cadmium (Cd), copper (Cu), nickel (Ni), lead (Pb) and Zinc (Zn) at disposal sites, at concentrations higher than the control sites. These metal contaminants cause physiological disorders in the soil and alter soil chemistry. Over the years, they bring significant damage to soil quality, fertility, and productivity. Such solid waste pollutants bringing change in physio-chemical characteristics of soil, also hamper the diversity of vegetation directly linked to those characteristics. Studies have also suggested enhancement of the organic matter content in the soil as one reason stimulating this change (Mouhoun-Chouaki, Derridj, Tazdaït, & Salah-Tazdaït, 2019). Open dumping waste sites and their adjoining areas are reported to show a lesser diversified variety of plants than their control (Ali, Pervaiz, Afzal, Hamid, & Yasmin, 2014). Loss of vegetation is also a hazard associated with waste dump sites (Yadav, Kumar, & Singh, 2018).

Researchers have indicated that heavy metals like lead, cadmium, chromium, zinc, arsenic, and mercury released from the stack emissions of solid waste incinerators, on the atmospheric deposition, is an important source of heavy metal pollution in soils adjacent to incinerator facilities (Zhong, et al., 2019). The mechanism is popular in treating hospital waste. With the expanding health industry in developing nations, the opportunities for practising incineration have also increased. Generally, incineration doesn't completely treat the metallic components of the waste introduced but rather collect the heavy metals into the bottom ash. This bottom ash, if uncontrollably disposed in open pits adjoining to the incineration facility or distributed on the surface of open dumps, can cause serious soil contamination in nearby areas. These bottom ash metals can percolate into the strata of the soil, cause groundwater contamination, and make the path for their runoffs into the water bodies. It ultimately enters the food chain through inhalation of heavy metal-laden dust from the dumpsites, bioaccumulation in plants and animals straying to the dump sites, and consumption from water sources polluted with the same (Adama, Esena, Fosu-Mensah, & Yirenya-Tawiah, 2016).

Composting, although considered as an appropriate mechanism for handling biodegradable waste, is responsible for causing secondary environmental pollution (Mandpe et al., 2019b). Compost generated from the mechanism, often used as a safer and more stable soil fertilizer and source of nitrogen, is responsible for causing nitrogen losses through the emissions of ammonia (NH_3) and nitrous oxide (N_2O). The leachate discharged during the process is considered as a source of soil pollution (Yang, et al., 2019). Thus, diverse methods of solid waste disposal contribute to soil contamination in varying ways (Mandpe et al., 2021a).

9.5.1 Composting linked Contagion

Composting can be simply defined as a cost-effective method of decomposing organic waste in the aerobic condition (Manyapu et al., 2019). However, when composting is performed on a bigger scale, it may pose a threat to environmental sustainability. Composting plants process waste and residual materials having a high loading of potentially pathogenic microorganisms such as viruses and bacteria, which makes the plant a significant source of bioaerosols. Bioaerosols resulting from the compost consist of live microorganisms as well as components of dead microorganisms and some inert particles. Bacterial peptidoglycans, cell wall glucans of molds, and airborne endotoxins constitute the components of dead microorganisms in the bioaerosols (Seidl 1995). Environmental conditions such as temperature, humidity, sun, wind, etc., as well as season, time of the day and distance from the source of aerosol play an important role in determining the levels of microorganisms in bioaerosol (Tong and Lighthart 1997; Folmsbee and Strevett 1999; Jones and Harrison 2004).

Bioaerosols originate mostly from molds, Gram negative bacteria, thermophilic actinomycetes, and viruses (Lacey and Crook 1988; Malmberg et al. 1993; Herr et al. 2003; Carvalheiro et al. 1995). There are a large number of factors influencing the emission and dispersal rate of bioaerosol. These factors include the nature of composted material, composting method, process equipment used, meteorological, topographical, and geographical features. (Marchand et al. 1995; Folmsbee and Strevett 1999; Reinthaler et al. 1997; Hryhorczuk et al. 2001; Taha et al. 2005).

Studies have shown that bacterial aerosol emission from composting plants is hazardous (Herr et al. 2003; Bünger et al. 2000). Bioaerosols in the occupational environment are also known to cause several health effects including infectious diseases, acute toxic effects, allergies, upper airway inflammation and cancer (Wouters et al. 2002; Douwes et al. 2000; Heldal et al. 2003a,b; Douwes et al. 2003).

9.5.2 Dumping Associated Contagion

Open dumping is the process of disposing solid waste at a land disposal site. This process does not protect the environment as the dumped waste is susceptible to open burning, and exposed to the several other elements, vectors, and scavengers in the ecosystem.

The size of the landfill depends on the quantity of waste generated in a country. No matter how small a landfill is, it significantly affects the lives of people working on them or living within 10km. These open dumping sites pose a serious threat to human health and water supplies, and all are located within or close to fast-growing cities in poor or low-income countries.

Disposal of solid waste spreads air, water, and soil pollution, which provides breeding ground to the biological vectors such as rodents, flies, insects, and pests. These vectors aid in spreading several diseases such as dysentery, worm infection, diarrhoea, food poisoning, cholera, dengue fever, leptospirosis, and bacterial infection (Pradyumna, 2013). It has been observed that the municipal drinking water of contaminated wells by waste disposal site even have an adverse effect on spontaneous abortions, congenital disabilities and other children heath concerns, and leukaemia (Wrensch et al., 1990; Wrensch et al., 1992).

People working in dumpsites are more prone to experience these diseases. To sum up this, the most common human health concerns arising from waste dumping are diseases related to gastrointestinal, respiratory & genetic systems and dermatological issues.

The waste dumps around the globe are growing in size and are posing a threat to the health of over 60 million people and are emerging as one of the sources of rivers, lakes, and ocean pollution (Vidal, 2014). As the waste dumped in these sites is not aided by mechanical processes such as compaction, the waste remains susceptible to open fire and the action of the weather. There is no engineering aid to leachate management or gas collection. This makes the open dumpsites to serve as pollution and health time bombs as these sites in addition to human health also distort the environment by groundwater and surface water pollution, soil contamination from potentially toxic elements, air pollution from open burning of waste, underground fires fueled by landfill gas and gas leakage.

9.5.3 Landfill site's Contagion

The landfill is the most used method of disposing of waste by burying it inside the land. This method is comparatively economical and popular across the globe, specifically in developing countries. However, this method serves as the major source of water, air, and land pollution and also plays a role in the spread of diseases among humans. This stands true, particularly for the people who reside near landfill sites and are worst affected (Mandpe et al., 2021c).

In addition to the leachate, gases released from landfill sites also act as a factor in degrading the environment and paving the way to the hazardous effects on human health. Local people are exposed by these gases via inhalation of air-born emissions or dust. The type of waste and phase of degradation determines the composition of gases. Still, the major gases released from the landfill are generally methane, carbon dioxide, oxides of nitrogen and Sulphur, chlorides of hydrogen, etc. (Maheshwari et al., 2015).

Exposure to nitrogen dioxide and Sulphur dioxide can cause inflammation & bronco constriction and may affect the immune system of the body. Hydrogen fluoride and hydrogen chloride serve as an irritant to mucosa membrane and tend to get deposited in the nostrils and upper respiratory tract leading to cough, chest tightness & breathlessness. (EPAQS, 2009).

There have been numerous incidents of health problems among children like eye irritation, skin rashes, abdominal pain, hypersensitivity, learning problems, incontinence, and seizures who lived close to landfill sites as compared to the children in areas away from landfill sites (Clark, 1982).

Studies have shown that from landfill sites a wide spectrum of degradation products are released into the environment in the form of gases like carbon dioxide, traces of hydrogen sulfide, VOCs, methane, and metal vapours (Zmirou et al., 1994; Ward et al., 1996; Hamar et al., 1996). The waste released may even cause cancer and stillbirth at times (Vrijheid, 2000; Goldberg et al., 1995). Cadmium (IARC, 1993 & 1994) is carcinogenic to humans and classified under Group 1, whereas styrene & lead (IARC, 1987a), formaldehyde (IARC, 1995) is also carcinogenic to human, classified under Group 2B & Group 2A respectively.

Studies have also shown the presence of Volatile organic compounds (VOC) in the blood of the population living near a landfill. The population was also observed to have an unusual situation of illness that they believe to be attributed to VOC exposure (Hamar et al., 1996). Exposure to VOCs, such as benzene releasing from the landfill site, is also known to cause leukaemia (IARC, 1987b).

Volatile organic compounds such as arsine and stibine are also generated from the halogenated or aromatic hydrocarbons present in the landfills. These compounds are acutely toxic, but in human's arsine gets converted to arsenic (WHO, 2000). In addition to the waste, pesticides which are commonly used on landfill sites also shows an association with brain cancer (Bohnen and Kurland, 1995)

9.5.4 Incineration linked Contagion

The waste-incineration approach have been in general underwent quite a change in the last few years. At present, the application of this approach is also in the context of reducing the waste volume and also the toxicity of the waste generated. However, these activities also affect the quantity and type of wastes incinerated and emissions generated.

With the advancement in the approach, incinerators are no longer fed with a heterogeneous mixture of unprocessed municipal solid waste and lower-volume waste streams from different sources. Also, the combustion process today is more regulated concerning temperature and oxygen control. Thus, reducing the chances of incomplete combustion. However, considerable emissions of air pollutants and uncontrolled handling of the ash residue are still a problem (Waste Incineration and Public Health, 2000).

In addition to the major combination product of combustion, i.e., Carbon dioxide and water vapour, by-products, such as residual ash, fly ash, and trace organic and inorganic compounds in the exhaust gases are also emitted. The composition of these incinerated waste streams determines the composition of emissions by the combustion process and reactions occurring in the waste gases after combustion. When the input wastes may have higher concentrations of nonfuel components (including metals, chlorine, sulfur, and nitrogen compounds) than fuels or biomass, there is a more solid residue (bottom ash and fly ash), and the concentrations of trace compounds in the waste gases tend to be higher than for combustion with an equal heat output from fuel or biomass (Waste Incineration and Public Health, 2000).

As discussed in the earlier sections, metals like mercury and cadmium remain unaffected by the combustion process and are released into the environment. These are precipitated from the atmosphere and reach the water bodies in the form of rain. This contaminated water upon consumption can significantly affect human health. The same goes for PICs (e.g., methane, ethane, acetylene, and benzene), which upon interaction with humans, may cause serious ailments.

References

Abdel-Shafy, H., & Mansour, M. (2018). Solid waste issue: Sources, composition, disposal, recycling, and valorization. Egyptian Journal of Petroleum , 275-1290.

Adama, M., Esena, R., Fosu-Mensah, B., & Yirenya-Tawiah, D. (2016). Heavy Metal Contamination of Soils around a Hospital Waste Incinerator Bottom Ash Dumps Site. Journal of Environmental and Public Health , 6.

Advanced Disposal. (2020). LEARN ABOUT LANDFILLS. Retrieved from https://www.advanceddisposal.com/: https://www.advanceddisposal.com/for-mother-earth/education-zone/learn-about-landfills.aspx

Ahmed, M. A., Diaz, C., Faaij, A., Gao, Q., Hashimoto, S., Mareckov, K., Zhang, T. (2007). Waste Management. Cambridge, United Kingdom and New York, NY, USA: Cambridge University Press.

Ahluwalia, I.J. and Pat, U. (2018). Solid Waste Management in India: An Assessment of Resource Recovery and Environmental Impact. Retrieved from https://think-asia.org/bitstream/handle/11540/8143/Working_Paper_356.pdf?sequence=1

Al Sabahi E, Rahim S.A., Wan Zuhairi W.Y., Al Nozaily F., Alshaebi F., 'The Characteristics of Leachate and Groundwater Pollution at Municipal Solid Waste Landfill of Ibb City, Yemen, Amer J Environ Sci 2009; 5 (3): 256-266.

Ali, M. H., & Abdel-Satar, A. M. (2005). Studies of some heavy metals in water, sediment, fish and fish diets in some fish farms in El-Fayoum province. Egypt. J. Aquat. Res., 31, 261-273.

Ali, S. M., Pervaiz, A., Afzal, B., Hamid, N., & Yasmin, A. (2014). Open dumping of municipal solid waste and its hazardous impacts on soil and vegetation diversity at waste dumping sites of Islamabad city. Journal of King Saud University-Science , 59-65.

Aliyu, B. N. (2008): Municipal Solid Waste Characteristics in three Residential Zones of Kano Metropolis: Implications for Management in the Book of Proceedings, 50th Annual Conference of the Association of Nigeria Geographers Uni. of Calabar, August 25th – 28th.

Allen, R. J., Brenniman, G. R., & Darling, C. (2012). Air Pollution Emissions from the Incineration of Hospital Waste. Journal of the Air Pollution Control Association, 829-831.

Bausback, B. (2016, April 27). The 3 Most Common Landfill Problems & Solutions. Retrieved from https://www.hcr-llc.com: https://www.hcr-llc.com/blog/the-3-most-common-landfill-problems-solutions

Bahnasawy, M., Khidr A., & Dheina, N. (2011). Assessment of heavy metal concentrations in water, plankton, and fish of Lake Manzala, Egypt. Turk. J. Zool., 35(2), 271-280.

Bohnen NI and Kurland LT. (1995). Brain tumor and exposure to pesticides in humans: a review of the epidemiologic data. J Neurol Sci. 132:110–121.

Bünger, J., Antlauf, Lammers, M., Schulz, T.G., Westphal, G., Müller, M., Ruhnau, P. and Hallier, E. (2000) Health complaints and immunological markers of exposure to bioaerosols among biowaste collectors and compost workers. Occup Environ Med 57, 458–464.

Carvalheiro, M.F., Peterson, Y., Rubenowitz, E. and Rylander, R. (1995) Bronchial reactivity and work, related symptoms in farmers. Am J Ind Med 27, 65–74.

Clark CS, Meyer CR, Balistreri WF, Gartside PS, Elia VJ, Majeti VA and Specker B. (1982). An environmental health survey of drinking water contamination by leachate from a pesticide waste dump in Hardeman County, Tennessee. Arch Environ Health. 1982 Jan-Feb;37(1):9–18

Conseve Energy Future. (2020, June 11). What is Open Burning? Retrieved from https://www.conserve-energy-future.com: https://www.conserve-energy- future.com/ effects-solutions -open-burning.php

Department of Primary Industries and Regional Development's Agriculture and Food division. (2018, July 23). Composting to avoid methane production. Retrieved July 15, 2020, from https://www.agric.wa.gov.au: https://www.agric.wa.gov.au/climate-change/composting-avoid-methane-production#:~:text=Composting%20is%20an%20aerobic%20process,to%20global%20greenhouse%20gas%20emissions.

Dhar, H., Kumar, S., Kumar, R., 2017. A review on organic waste to energy systems in India. Bioresour. Technol. 245, 1229–1237. https://doi.org/10.1016/j.biortech.2017.08.159

Douwes, J., Wouters, I., Dubbeld, H., van Zwieten, L., Steerenberg, P., Doekes, G. and Heederik, D. (2000) Upper airway inflammation assessed by nasal lavage in compost workers: a relation with bio-aerosol exposure. Am J Ind Med 37, 459–468.

Douwes, J., Thorne, P., Pearce, N. and Heederik, D. (2003) Bioaerosol health effects and exposure assessment: progress and prospects. Ann Occup Hyg 47, 187–200.

Eitzer, B. D. (1995). Emissions of Volatile Organic Chemicals from Municipal Solid Waste Composting Facilities. Environmental Science & Technology, 896-902.

Environment Victoria. (2013, June 16). The problem with landfill. Retrieved June 13, 2020, from https://environmentvictoria.org.au: https://environmentvictoria.org.au/resource/problem-landfill/

Environmental Protection Agency. (2017, February 14). Greenhouse Gas Emissions. Retrieved from https://www.epa.gov/ghgemissions/understanding-global-warming potentials#:~:text=Methane%20(CH4)%20is%20estimated,uses%20a%20different%20value.).&text=Nitrous % 20Oxide%20(N2O, for%20a%20100%2Dyear%20timescale.

Environmental Protection Agency. (2020, March 26). What is a Municipal Solid Waste Landfill? Retrieved from https://www.epa.gov: https://www.epa.gov/landfills/municipal-solid-waste-landfills

EPQS (Expert Pannel on Air quality standards), (2009). Adendum to Guidelines for Halogens and Hydrogen Halides in Ambient Air. London; The stationary office.

Federal Facilities Environmental Stewardship & Compliance Assistance Center. (2017, July 05). Open Dumping. Retrieved from https://www.fedcenter.gov: https://www.fedcenter.gov/assistance/facilitytour/solid/dumping/#:~:text=An%20open%20dumping%20is%20defined,elements%2C%20vectors%2C%20and%20scavengers.&text=An%20officially%20designated%20Sanitary%20Waste%20Landfill%20is%20not%20an%20open%20dump.

Fischer, C., Maurice, C., & Lagerkvist, A. (1999). Gas Emission from Landfills. Stockholm: Swedish Environmental Protection Agency.

Folmsbee, M. and Strevett, K. (1999) Bioaerosol concentration at an outdoor composting center. J Air Waste Manage Assoc 49, 554–561.

Goldberg MS, al-Homsi N, Goulet L and Riberdy H. (1995). Incidence of cancer among persons living near a municipal solid waste landfill site in Montreal, Québec. Arch Environ Health. 1995 Nov-Dec;50(6):416–424.

Hadjibiros, K., Dermatas, D. and Laspidou, C.S. (2011). Municipal solid waste management and landfill site selection in Greece: irrationality versus efficiency. Journal of Global NEST, 13(2): 150-161

Hammer GB, McGeehin MA, Phifer BL and Ashley DL. (1996): Volatile organic compound testing of a population living near a hazardous waste site. J Expo Anal Environ Epidemiol. 1996 Apr-Jun; 6(2):247-55.

He, Y., Inamori, Y., Mizuochi, M., Kong, H., Iwami, N., & Sun, T. (2001). Nitrous Oxide Emissions from Aerated Composting of Organic Waste. Environmental Science & Technology, 2347–2351.

Heimlich, J. E. (n.d.). LANDFILL VERSUS DUMP: SOME FACTS. Retrieved from https://www.guyenterprisegy.com: https://www.guyenterprisegy.com/solidwaste/dumpvslan.htm Heldal, K.K., Halstensen, A.S., Thorn, J., Djupesland, P., Wouters, I., Eduard, W. and Halstensen, T.S. (2003a) Upper airway inflammation in waste handlers exposed to bioaerosols. Occup Environ Med 60, 444–450.

Herr, C.E.W., Zur Nieden, A., Jankofsky, M., Stilianakis, N.I., Boedeker, R.H. and Eikmann, T.F. (2003) Effects of bioaerosols polluted outdoor air on airways of residents: a cross sectional study. Occup Environ Med 60, 336–342.

Hryhorczuk, D., Curtis, L., Schleff, P., Chung, J., Rizzo, M., Lewis, C., Keys, N. and Moomey, M. (2001) Bioaerosols emission from a suburban yard waste composting facility. Ann Agric Environ Med 8, 177–185.

IARC (1987a): Monographs on the Evaluation of Carcinogenic Risks to Humans Some Metals and Metallic Compounds. Intl Agency Res Cancer, Lyon. 1987a; 23 Suppl. 7:325–438.

IARC (1987b): Monographs on the Evaluation of Carcinogenic Risks to Humans. Some Industrial Chemicals and Dyestuffs. Intl Agency Res Cancer, Lyon. 1987b; 29 Suppl. 7:93–148.

IARC (1993): Monographs on the Evaluation of Carcinogenic Risks to Humans. Beryllium, Cadmium, Mercury, and Exposures in the Glass Manufacturing Industry. Intl Agency Res Cancer, Lyon. 58:119–238.

IARC (1994): Monographs on the Evaluation of Carcinogenic Risks to Humans Some Industrial Chemicals. Intl Agency Res Cancer, Lyon. 60:233–320.

IARC (1995): Monographs on the Evaluation of Carcinogenic Risks to Humans. Wood Dust and Formaldehyde. Intl Agency Res Cancer, Lyon. 62:217–405.

Isaifan, R., & Alyasi, H. S. (2018). A Review on Pollution Emissions and Impact from Waste Treatment and Disposal Facilities. Journal of Environmental and Toxicological Studies.

Jain, C.K., Kumar, C.P. and Sharma, M.K. (2004). Irrigation water quality of Ghataprabha Command Area, Karnataka. Water Pollution, A.P.H. Publishing Corporation, New Delhi, 17: 145-158.

Jhamnani, B. and Singh, S.K. (2009). Ground water contamination due to Bhalaswa landfill Site in New Delhi. International Journal of Civil and Environmental Engineering, 1(3): 121-125.

Johnke, B. (n.d.). Emissions from Waste Incineration. Intergovernmental Panel on Climate Change.

Jones, A.M. and Harrison, R.M. (2004). The effects of meteorological factors on atmospheric bioaerosol concentrations – a review. Sci Total Environ 326, 151–180.

Kamboj, N. and Choudhary, M. (2013). Impact of solid waste disposal on ground water quality near Gazipur dumping site, Delhi, India. Journal of Applied and Natural Science, 5 (2): 306-312.

Kumar, S., Negi, S., Mandpe, A., Singh, R.V., Hussain, A., 2018. Rapid composting techniques in Indian context and utilization of black soldier fly for enhanced decomposition of biodegradable wastes - A comprehensive review. J. Environ. Manage. 227, 189–199. doi:10.1016/j.jenvman.2018.08.096

Lacey, J. and Crook, B. (1988) Fungal and actinomycetes spores as pollutants of the workplace and occupational allergens (review). Ann Occup Hyg 32, 515–533.

Landfill Methane Outreach Program (LMOP). (n.d.). Retrieved from https://www.epa.gov: https://www.epa.gov/lmop/basic-information-about-landfill-gas#:~:text=Methane%20Emissions%20from%20Landfills,of%20these%20emissions%20in%202018.

Lou, X. F., & Nair, J. (2009). The impact of landfilling and composting on greenhouse gas emissions – A review. Bioresource Technology, 3792-3798.

Maheshwari R, Gupta S and Das K. (2015). Impact of Landfill Waste on Health: An Overview. IOSR Journal of Environmental Science, Toxicology and Food Technology. 1(4):17-23.

Majumdar, D., & Ganguly, S. (2018). Air Pollution from Municipal Solid Waste Management Techniques. Waste Management and Resource Efficiency, 1115-1132.

Malmberg, P., Rask, Andersen, A. and Rosenhall, L. (1993) Exposure to microorganisms associated with allergic alveolitis and febrile reactions to mould dust in farmers. Chest 103, 1202–1209.

Mandpe, A., Kumar, S., Lakshmikanthan, P., Hettiarachchi, H., 2019a. Mining for recovery as an option for dumpsite rehabilitation: Case study from Nagpur, India. J. Environ. Eng. Sci. 15, 52–60. doi:10.1680/jenes.19.00021

Mandpe, A., Paliya, S., Kumar, S., Kumar, R., 2019b. Fly ash as an additive for enhancing microbial and enzymatic activities in in-vessel composting of organic wastes. Bioresour. Technol. 293, 122047. doi:10.1016/j.biortech.2019.122047

Mandpe, A., Kumari, S., Kumar, S., 2020a. Composting: A Sustainable Route for Processing of Biodegradable Waste in India. Org. Waste Compost. through Nexus Think. 39–60. doi:10.1007/978-3-030-36283-6_3

Mandpe, A., Tyagi, L., Paliya, S., Chaudhry, S., Motghare, A., Kumar, S., 2021a. Rapid-in-house Composting of Organic Solid Wastes with Fly Ash Supplementation: Performance Evaluation at Thermophilic Exposures. Bioresour. Technol. 337, 125386. doi:10.1016/j.biortech.2021.125386

Mandpe, A., Pratap, V., Lakshmikanthan P., 2021c. Landfill mining. In Current Developments in Biotechnology and Bioengineering (pp. 197-229). Elsevier. doi: 10.1016/B978-0-12-821009-3.00010-5

Manyapu, V., Mandpe, A., Kumar, S., 2019. Corrigendum to "Synergistic effect of fly ash in in-vessel composting of biomass and kitchen waste" [Bioresour. Technol. 251 (2018) 114–120] (S0960852417321740)(10.1016/j.biortech.2017.12.039). Bioresour. Technol. 277, 225. doi:10.1016/j.biortech.2018.12.031

Marchand, G., Lavoie, J. and Lazure, J. (1995) Evaluation of bioaerosols in a municipal solid waste recycling and composting plant. J Air Waste Manage Assoc 45, 778–781.

Mavropoulos, A. (2013). The Tragic Case of Dumpsites. Auerspergstrasse: International Solid Waste Association.

Mohareb, E., MacLean, H., & Kennedy, C. (2011). Greenhouse Gas Emissions from Waste – An Assessment of Quantification Methods. Journal of the Air & Waste Management Association, 480-493.

Mouhoun-Chouaki, S., Derridj, A., Tazdaït, D., & Salah-Tazdaït, R. (2019). A Study of the Impact of Municipal Solid Waste on Some Soil Physicochemical Properties: The Case of the Landfill of Ain-El-Hammam Municipality, Algeria. Applied and Environmental Soil Science, 8.

Negi, S., Mandpe, A., Hussain, A., Kumar, S., 2020. Collegial effect of maggots larvae and garbage enzyme in rapid composting of food waste with wheat straw or biomass waste. J. Clean. Prod. 258. doi:10.1016/j.jclepro.2020.120854

Nzeadibe, T. C. (2009): Solid Waste Reforms and Informal Recycling in Enugu Urban Area, Nigeria. Habitat International. 33 (1)93-99

Ocheri, M. I.; Mile, Y. and Oklo, A. (2008): "Nitrate Contamination of Shallow Wells in Makurdi Urban Area of Benue State" in the Book of Proceedings, 50th Annual Conference of the Association of Nigeria Geographers Uni. of Calabar, August 25th – 28th.

Paliya, S., Mandpe, A., Kumar, S., Kumar, M.S., 2019. Enhanced nodulation and higher germination using sludge ash as a carrier for biofertilizer production. J. Environ. Manage. 250, 109523. doi:10.1016/j.jenvman.2019.109523

Pillai, S., Peter, A. E., M, S. B., & S., S. (2014). Soil Pollution near a Municipal Solid Waste. International Conference on Biological, Civil and Environmental Engineering , 148-152.

Pradyumna A. (2013). Understanding the health risks of solid waste management practices applying evidences to Bangalore's context. SOCHARA, Bangalore. January, 2013. http://www.academia.edu/5222236/Understanding_the_health_risks_of_solid_waste_management_practices_applying_evidence_to_Bangalores_context

Rajkumar, N., Subramani, T. and Elango, L. (2010). Groundwater contamination due to municipal solid waste disposal – A GIS based study in erode city. International Journal of Environmental Sciences, 1: 39-55.

Reinthaler, F.F., Marth, E., Eibel, U., Enayat, U., Feenstra, O., Friedl, H., Köck, M., Pichler Semmelrock, F.P. et al. (1997) The assessment of airborne microorganisms in large-scale composting facilities and their immediate surroundings. Aerobiologia 13, 167–175.

Sabahi, E.A., Rahim, S.A., Zuhairi, W.Y.W. and Nozaily, F.A. (2009). Assessment of groundwater pollution at municipal solid waste of Ibb landfill in Yemen. Bulletin of the Geological Society of Malaysia, 55: 21–26.

Sadek, S. and El-Fadel, M. (2000). The Normandy landfill: A Case Study in solid waste management. Journal of Natural Resources and Life Sciences Education, 29: 155-161.

Science.jrank. (2020). Incineration: Emissions of Polltants. Retrieved June 14, 2020, from https://science.jrank.org: https://science.jrank.org/pages/3548/Incineration-Emissions-pollutants.html

Seidl, H.P. (1995) Mikrobiologie des Abfalls. In Keimbelastung in der Abfallwirtschaft: Tagung 26/4/1995 ed. Mücke, W., Seidl, H.P., Rakoski, A.H., Eckrich, C., Emmerling, G., Pipke, R. and Wimmer, M. pp. 5–30. München, Germany: Klinikum rechts der Isar der Technischen Universität München

Sharma, B. K., & Chandel, M. K. (2018). Emissions Associated with the Transportation of Municipal Solid Waste in Mumbai. Advances in Waste Management, 437-444.

Smet, E., Langenhove, H. V., & Bo, I. D. (1999). The emission of volatile compounds during the aerobic and the combined anaerobic/aerobic composting of biowaste. Atmospheric Environment, 1295-1303.

SN, C. and Chadar, K. (2017). Solid Waste Pollution: A Hazard to Environment. Recent Advances in Petrochemical Science, 41-43.

Stewart, E. S., & Lemieux, P. M. (2003). Emissions from the incineration of electronics industry waste. IEEE.

Taddese, S. (2019). Municipal Waste Disposal on Soil Quality. A Review. Acta Scientific Agriculture , 09-15.

Taha, M.P.M., Pollard, S.J.T., Sarkar, U. and Longhurst, P. (2005) Estimating fugitive bioaerosol releases from static compost windrows: feasibility of a portable wind tunnel approach. Waste Manag 25, 445–450.

Tan, R. B., & Khoo, H. H. (2012). Impact Assessment of Waste Management Options in Singapore. Journal of the Air & Waste Management Association, 244-254.

Tong, Y. and Lighthart, B. (1997) Solar radiation has a lethal effect on natural populations of culturable outdoor atmospheric bacteria. Atmos Environ 31, 897–900.

The World Bank, Solid Waste Management. (2018). Retrieved from http://www.worldbank.org/en/topic/urbandevelopment/brief/solid-waste-management

Uting, p. B.; Andrew-essien, E.; Akintoye, A.O. and Upla, P.I. (2008): "Assessing the Temporal and Spatial patterns in flow regime in Aya River, South Eastern Nigeria". Global Journal of Environmental Sciences, 7(1), 2008.

Vidal, J. Smelly, contaminated, full of disease: the world's open dumps are growing. The Guardian [Internet]. 2014 October 6. Available from https://www.theguardian.com/global-development/2014/oct/06/smelly-contaminated-disease-worlds-open-dumps

Vrijheid M. (2000). Health effects of residence near hazardous waste landfill sites: A review of epidemiologic literature. Environ. Health Perspect. 108, 101–112.

Ward RS, Williams GM and Hills CC. (1996): Changes in major and trace components of landfill gas during subsurface migration. Waste Manage Res. 14:243–261.

Waste Decomposition and Degradation of Waste in Landfills. (2019, April 11). Retrieved June 06, 2020, from https://landfill-site.com: https://landfill-site.com/waste-decomposition.html Waste Incineration And Public Health. [online] National Academy Press Washington, D.C. Available at: <https://www.nap.edu/read/5803/chapter/4> Weitz, K. A., Thorneloe, S. A., Nishtala, S. R., Yarkosky, S., & Zannes, M. (2002). The Impact of Municipal Solid Waste Management on Greenhouse Gas Emissions in the United States. Air & Waste Management Association, 1000-1011.

Weitz, K., Barlaz, M., Ranjithan, R., Brill, D., Thorneloe, S., & Ham, R. (1999). Life Cycle Management of Municipal Solid Waste. The International Journal of Life Cycle Assessment , 195-201.

Wouters, I.M., Hilhorst, S.K.M., Kleppe, P., Doekes, G., Douwes, J., Peretz, C. and Heederik, D. (2002) Upper airway inflammation and respiratory symptoms in domestic waste collectors. Occup Environ Med 59, 106–112.

Wrensch M, Swan S, Murphy PJ, Lipscomb J, Claxton K, Epstein D and Neutra R. (1990). Hydrogeologic assessment of exposure to solvent-contaminated drinking water: pregnancy outcomes in relation to exposure. Arch Environ Health 45:210—216.

Wrensch M, Swan SH, Lipscomb J, Epstein DM, Neutra RR and Fenster L. (1992). Spontaneous abortions and birth defects related to tap and bottled water use, San Jose, California, 1980-1985. Epidemiology. Mar;3 (2):98–103.

Yadav, H., Kumar, P., & Singh, V. P. (2018). Hazards from the Municipal Solid Waste Dumpsites: A Review. Proceedings of the 1st International Conference on Sustainable Waste Management through Design, 336-342.

Yang, X., Liu, E., Zhu, X., Wang, H., Liu, H., Liu, X., et al. (2019). Impact of Composting Methods on Nitrogen Retention and Losses during Dairy Manure Composting. International Journal of Environmental Research and Public Health, 17.

Zhang, H., Schuchardt, F., Li, G., Yang, J., & Yang, Q. (2013). Emission of volatile sulphur compounds during composting of municipal solid waste (MSW). Waste Management, 957-963

Zhong, S., Wei, Z., Zhang, L., Lei, C., Gao, H., Feng, J., et al. (2019). Spatial distribution of heavy metals in soil around municipal solid waste incinerators at coastal plain and inland mountainous region in China. IOP Conference Series: Earth and Environmental Science, 12.

Zmirou D, Deloraine A, Saviuc P, Tillier C, Boucharlat A and Maury N. (1994): Short-term health effects of an industrial toxic waste landfill: a retrospective follow-up study in Montchanin, France. Arch Environ Health. (4):228– 238

10

Environmental Education and Communication as a Foundation in Sustainable Development

***Prachi Goyal*[a,b] *and Sonam Paliyaa*[ac*]**

[a]*Academy of Scientific and Innovative Research (AcSIR) Ghaziabad- 201 002, India*
[b]*CSIR-National Environmental Engineering Research Institute, Delhi Zonal Centre, A-93/94, Naraina Industrial, Area Phase-1, New Delhi-110028, India*
[c]*CSIR-National Environmental Engineering Research Institute (CSIR-NEERI), Nehru Marg, Nagpur-440 020, India*

ABSTRACT

Environment Education & Communication is an interdisciplinary, innovative and dynamic concept that goes beyond raising awareness to help individuals and groups, acquire knowledge and skills to change behaviours towards specific environmental issues in order to attain environmental dimension of sustainable development. The prime goal is to express the deep essence of the new form of human development through education and communication that intends to improve the quality of life and fostering a healthy and safe environment to the future generations. There are many different stakeholders involved in sustainable development such as governments, intergovernmental bodies, the mass media, the civil society, non-governmental organizations, the private sector and formal education institutions (i.e. schools, universities and research and training institutes) whose participation and concerted efforts are essential for long term developments. Recognizing the importance and need to safeguard natural resources, nations also feel an increasing need to cooperate and coordinate with global partners for promoting environmental regulations. The chapter, hereby, focuses on strategic use of education and communication in synergy with stakeholders and global partners for effective policy making, active participation and implementation that can lead to long term transformations geared towards environmental sustainability

Keywords: Environment, Education, Communication, Sustainability, Public Participation, Climate Change

10.1 Introduction

The consistent process of rapid urbanization and expeditious developmental activities have severely impacted the environmental components and thereby the quality of life. With limited natural resources, individuals and societies have to learn to bring in attitudinal changes in order to accomplish environment sustainability (UNESCO). From the Book "**Our Common Future**" (Brundtland Report), the concept of **Sustainable Development** can be understood as the "development that meets the needs of the present, but does not threaten the ability of future generations to meet their own needs". The definition and concept became most widely-spread after its universal acceptance at UN Conference on Environment and Development (UNCED) in 1992 in Rio de Janerio (Alexandar, 2014). Talking about the major challenges, environmental degradation, pollution, climatic phenomenon, natural calamities, poverty and social inequalities are some of the areas that needs to be dealt with concern. The interrelatedness and interdependencies of human about activities on natural resources have posed a critical threat to the environment. Lack of awareness, knowledge, skills and casual attitudes have further aggravated the problem. Thus, in view of diminishing the use of natural resources and safeguarding the environment, **Environment Education and Communication (EE&C)** has emerged as an essential tool. It has also gained importance as one of the top most agendas nationally as well as globally for achieving transformation from self-centered use of environmental components to its sustainable use (Bhat *et al.*, 2017).

The strategy framework in order to attain a sustainable future should be designed on following objectives:

1. The information that will be received by the audience **(WHO)**
2. The information content **(WHAT)**
3. The methods of communication **(HOW)**
4. The framework incorporating measures to promote change **(ACTION)** (USEPA).

10.2 Environment Education and Communication (EE&C)

Environment Education is a process that aims at preparing its citizen with technical knowledge, skills and strategies for solving environment related issues. **Communication** is defined in terms of strategic use of communication processes and media products to understand key environmental factors, their interdependencies and to respond to the problems in a competent way. Environment Communication is based on the principle of **3R**, i.e. **Responsible, Rational and Relevant** (Paulo). The goal of promoting environmental education and communication is to instill in learners the knowledge and positive attitude towards the environment, competency in citizen action skills and a sense of empowerment (Brian A. Day, 2000). Simultaneously, it plays a vital role in bringing masses together for a common cause of protecting environment by sharing knowledge, skills, ideas and beliefs. Thus, achieving sustainable development calls for

- Awareness campaigns aimed at developing understanding about the environment
- Recognizing how man acts as an integral component of the environment
- Adopting measures and strategies for resource conservation (Festus and Ogoegbunam, 2012).

The management of solid waste, air pollution, water pollution including other environmental components has become a very big challenge in recent times.

Steady progress has been made to reduce pollution at semi-urban levels, such as those generating from polluted air and waterways. For example, the use of improved chulhas have not only reduced household air pollution but also the associated respiratory disorders, and access to improved sanitation facilities has reduced water borne diseases. Solid waste management has achieved a great success due to the implementation of education at various levels through workshops, exhibitions, lecture series, panel discussions, group/ community meetings *etc*. Also, the attitudinal changes adopted by different stakeholders involving scientific waste disposal by 4R concept- reduce, reuse, recycle and recovery of waste, storage and segregation at the source have gained significant importance. There are different campaigns organized at individual, national and/or international levels such as Chipko Movement, Swacch Bharat Mission, International Solar Alliance, Appiko Movement, Narmada Bachao Andolan *etc.* that have resulted in open defecation free country, promoted the need of afforestation, more reliability on renewable sources and the need for conservation and protection of nature and its associated resources.

Environment Education and Communication has turned up as a connecting link between human and environment, thereby, explaining the interdependency and interrelatedness between the two and how the changes in lifestyle, behavioral and cooperation amongst different interest groups can lead to a sustainable environment. Consequently, there are different stakeholders who have adopted several moves and initiatives to help individuals in building a new perspective towards environmental issues. Some of these are described in the subsequent section.

Eco-Schools Program: The program was developed in 1994 at UN (Rio) Conference on Environment and Development. The program was initiated with a view to extend learning beyond the classroom and involve young people in finding solutions to environmental and sustainable development changes at local level. Northern Ireland was one of the first countries to indulge and participate in the same. Over the years, research has shown that the schools and pupils care for the environment. The pupils turned out to be the keen litter collectors and encouraged the school to ban the use of plastic straws (Ruth Van Ry).

Science Express Climate Action Special (SECAS), India: In 2007, Department of Science & Technology (DST) developed a 16-coach air-conditioned train. The unique mobile science has been used to impart climate change literacy on wheels. Since its inception, it has travelled across India to explore variety of environmentally themed topics. It has proved to be an innovative method to educate people, especially students. Through a comprehensive evaluation, research has also shown that the initiative has led to an increased understanding of the measures that can be opted for climate change and the willingness to act (Sharma).

Graded Response Action Plan: It is a plan notified by government for Delhi and NCR with a view to reduce urban air pollution by adopting various control options under different AQI categories. The initiative enabled people to think critically and make informed and sound decisions such as carpooling, use of public transport *etc* (MoEF&CC, 2017).

International Solar Alliance: Recognizing the increasing demands of energy and climate change, France and India launched ISA in 2015. The project was initiated with the aim of solar energy deployment in 121 countries (ISA, 2020).

10.3 Implementation of Environmental Education in Educational System

India is one of the few countries where Environment Education became an integral component at all levels of formal education. It was the 42nd amendment in 1976 after which India incorporated environmental concerns in the constitution. Environmental Education has been introduced as a component at all levels of education following the directive of the Honorable Supreme Court on 18 December 2003 (Sonowal, 2009; Bhat *et al.*, 2017). **Environmental Education (EE**) is an interdisciplinary concept that involves education, sensitization towards environment, technical know-how, behavioral and/or attitudinal changes for recognizing the services that nature provides and understanding the interdependency of humans on environment (Khademi-Vidra, 2017; Malley, 2020). It also has an eminent role in the realization of SDGs (including poverty, education among others) that have gained global priorities and needs to be achieved by 2030. The basic components of environmental education are

1. **Knowledge:** In order to make responsible environmental decisions, one should understand following concepts
    - Natural environmental phenomenon
    - Effects of anthropogenic activities on the environment
    - Environmental laws and judiciary
    - Environment related issues such as climate change, pollution, land degradation *etc*.
    - Interdependencies and Interrelatedness between human and environment
    - Environment Protection measures, strategies and control options
2. **Skills:** There are six defined skills that work as prerequisites to identify and help resolve environmental challenges

- Communication
- Numeric
- Technical Knowledge
- Solutions to Problems
- Self-awareness
- Interpersonal Relations

2. **Attitude:** It includes the following aspects.
   - Concern for the environment
   - Respecting others views and opinions
   - Critical and independent thought on environmental issues
3. **Responsibility** towards environment
4. **Voluntary Participation** (USEPA; Gopal and Anand, 2008).

The basic objectives state that the aim of implementing environmental education is to impart critical thinking among individuals rather than advocating a particular view point. Also, enables citizens in recognizing environmental problems and technically solving them. Fig. 1 describes the schematic of the strategic framework that can be opted at school/ higher education levels in attaining environmental sustainability.

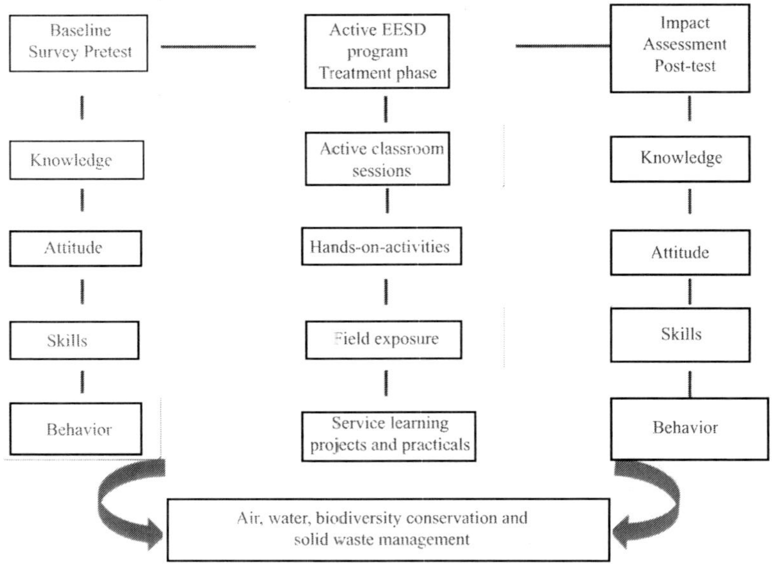

Fig. 1: A schematic of strategy design for Adopting Sustainable Development (Alexandar, 2014)

10.3.1 Implementation in School Education

In India, Environmental Education has been used as a tool for **Education for Sustainable Development (ESD)** at the school level. It makes individuals realize and recognize the interdependencies between human beings and various units of ecology. Indian government directed education departments to actively work on Environment Education. The main focus of imparting environmental education at all levels is student's exposure to the extremities of the real world. This, also enables them to identify, analyze and evaluate different environmental issues and make informed and responsible decisions with a goal of working in harmony with nature (Praveen and Nasreen; Sonowal, 2009; Bhat *et al.*, 2017)

Initially, only Uttarakhand included EE as a component of the curriculum until the class eighth. Following the National Curriculum for Elementary and Secondary Education- A Framework, formulated in 1988, first systematic and concentrated efforts were made to bring in EE as a part of school education at all levels (Sonowal, 2009). The major focus of NCF was to use EE as a tool in promoting logical and critical thinking as opposed to rotting memorizing the textbook content.

In light of this, NCERT suggested the following modes of a transaction of the curriculum at various levels of school education in India as described in Table 1.

Table 1: Transaction Modes for Environment Education at School Level

Classes	Age Group	Mode of Transaction
I and II	6-7	Different Interactive Activities
III to V	8-11	Environmental Studies (EVS)
VI to X	12-16	Infusion Approach
XI to XII	17-18	Project based study

Sharma; Praveen and Nasreen; NCF, 2005

- **Classes I and II (ages 6-7):** EE is activity oriented, *i.e.* it is integrated with the main subjects through activities. For example, students may study different colored flowers and plants by identifying them and involving in different related activities in the environment.
- **Classes III to V (ages 8-11):** EE is introduced as Environmental Studies (EVS), with a textbook named 'My Environment' that focuses on acquiring knowledge about the environment and its interrelationship with human beings.
- **Classes VI to X (ages 12-16):** Involves infusion model that integrates EE into other core subjects such as science, social studies, mathematics, language, *etc.* as the field itself is multidisciplinary.

- **Classes XI-XII (ages 17-18):** EE is included as a part of the student's interdisciplinary project in Elective or General Studies depending on the discipline opted by the student. EE has basically acquired through project based learning.

In addition to the above discussed framework followed in India, there are different curricula adopted by different countries at the school level. Some of these are

NatureWise, Netherland: NatureWise is a program supported by Dutch Forestry Service and Nature College Foundation that has been developed to (re)connect primary school children with nature and to simulate their care for nature by engaging them in outdoor experiences. The program designs the primary school curriculum followed for nature education. The framework is known as 3x3x3 Approach that includes following components.

- **Three Days:** Three days in three different seasons are planned with an official Forest Ranger for outdoor learning. The days are linked with some central theme and are elaborated with each day specific learning goals.
- **Three Experiential Domains:** Head, hands and heart are utilized to discover nature. The details of the same have been described in Fig. 2.

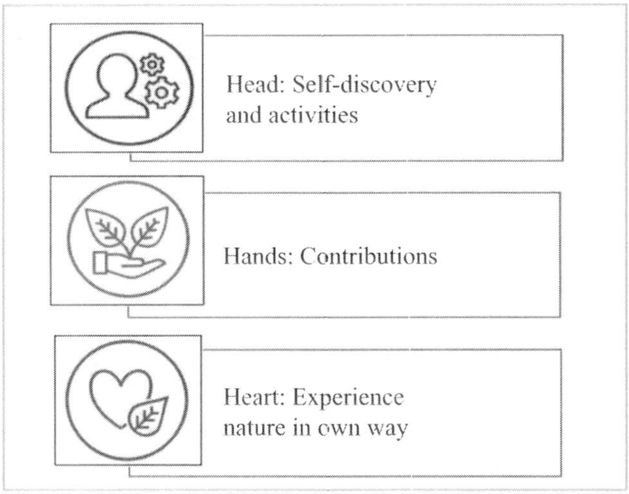

Fig. 2: Experiential Domains adopted at Nature Wise (Wals)

- **Three Chronological Phases:** School based preparatory, outdoor experiences, School-based reflection are the three chronological phases within the framework. Children engage with theme through outdoor experiences and explore nature in their own way. Further, they are evaluated based on their experiences through assignments or related activities (Wals)

Taiwan: The concept of environmental education gain widespread importance in Taiwan in 2001. The value on the importance gained further momentum when the discrete activities were incorporated into the national school curriculum. Initially, the integration was planned for grades 1 through 9 later it was extended till 12th grade. The themes and concepts pertaining to the framework are based on environmental ethics, sustainable development, climate change, disaster preparedness and sustainable use of energy and resources (Chang, 2016).

Based on the concentrated and continuous efforts by the Government and associated agencies in promoting Environment Education at the very basic school level, it can be easily deduced that Environment Education not only inspires to acquire responsible citizenship but also equips the future custodians of the environment to keep mother earth in safer hands. It also encourages us to make considerable efforts in protecting the same from any potential threats.

10.3.2 Implementation in Higher Education

In a bid to create awareness amongst youth about the sustainable environment for future generations, the Supreme Court passed a directive following Mehta's PIL in 1991 to make the subject of Environment Studies compulsory in the college curricula. Thereafter, the University Grants Commission (regulatory authority for higher education funding) instructed all the universities for including a six-month core module syllabus on Environmental Sciences in all the undergraduate courses. However, only a few universities included the same due to infrastructural upgrades and time-consuming curriculum development. Recently, it has been reported that 306 universities are yet to include EVS as a part of course work (*India Today*, 2017).

Following the Supreme Court Directive, UGC along with an expert Committee framed a module syllabus for Ability Enhancement Compulsory Courses (AECC- Environmental Studies) under Choice Based Credit System (CBCS) (*Asian Age*, 2017; Pradhan, 2019). The six month core module syllabus comprises of eight units, of which, unit one discusses about the fundamentals and basic concepts of environment studies, second and third unit provides details about ecosystem and renewable and non-renewable energy types, unit four discusses biodiversity and sustainable use of the components and resources derived from it, unit five and six highlights topics like environment pollution, policies for preservation, conservation of environment and abatement of pollution, unit seven is about the interrelatedness and interdependencies of human communities on environment. Unit eight focusses on the field visits and projects that aim at developing critical thinking and informed decision-making ability among the students. Thus, UGC after the Supreme Court's directive focused on strict implementation and enforcement of Environment Education as part of college course curricula to encourage, promote and inculcate a sense of environment conservation.

10.3.3 Educating Youth Group for Environmental Thinking

As environmental concerns escalate the globe, young perceptions and skills become indispensable for responsive changes and solutions to complex environmental problems. Since the youth population will have to live longer with the extended period of the consequences of current environmental decisions, their active role and participation play a vital role in overcoming the ecological challenges. Young people engage in a variety of innovative and creative tactics in their environmental advocacy- from painting murals to riding bicycles, from theatre arts to music to hip hops, from building websites to small enterprises. Their efforts range from local initiatives from cleaning neighbourhoods, parks, adopting environment friendly activities, preserving resources such as water and electricity to national and international awareness campaigns. Engaging youth in environment protection will not only bring about attitudinal and behavioural changes but will influence their families, parents, and relatives as well (Foundation Y.I.F., 2008; Ramchandra Acharya, 2013; Sayal *et al.*, 2016)

By changing lifestyle, inculcating green knowledge in the form of green tips, youth can contribute to promoting a cleaner, greener and sustainable environment. Young people also bring in a fresh perspective, critical thinking, energy, enthusiasm, drive and sense of what is possible to a wider extent. Some of the environmental and conservation strategies/ tips that young generations ought to bring in their daily routines are described below.

Going Green in House

- Turning off taps when not in use
- Closing running taps when not needed
- Saving electricity by switching off lights/fans *etc.* when not in use
- Relying more on natural renewable sources like solar power

Going Green with Recycling and Reusing

- Using local trash pickup service for recycling paper, glass, plastics and metals

- Using paper bags rather than plastic bags
- Choosing reusable bottles rather than plastic ones

Going Green with the Transport

- Carpooling
- Going for regular pollution checks
- Using public transport

Acknowledging the creative skills and mind sets that youth possess, it can be concluded that they can lead to invention and development of new forms of action and activism towards environmental issues. They play a pivotal role in critical thinking and making informed decisions regarding environmental concerns. Also, because of their longer life expectancy, they are expected to safeguard the environment for both the present and future generations (Ramchandra Acharya, 2013; Sayal *et al.*, 2016).

10.4 Challenges of Public Participation and Possible Solutions

The environmental concerns are peculiar of interest conflicts that shall not depend on the government's unilateral interventions. The coordination of several stakeholders (such as individuals, interest groups, communities) shall be realized through their active participation and practice in both environmental administration and judicial processes. The incorporation and integration of public participation in environmental issues lead to better understanding and acknowledgment of policies of the government or organizations. It, therefore, plays a crucial role in bringing the interests of diverse entities together (USEPA, a; Hu and Tang, 2014) environmental protection requires public participation. In order to realize the ideals of public participation in environmental protection, we need to enhance the democratization of national administrative agencies, guarantee the civil environmental rights, balance the environmental interests of different groups in light of environmental justice, promote public environmental awareness and cultivate the self-governing spirit in environmental protection. As there are definitely many problems in the public participation of environmental protection, we need to strengthen education in environmental protection, promote public participation in environmental protection and provide support to the civil environmental organizations so as to realize public participation in environmental protection. Public participation is basically a dynamic process that involves, informs and consults the public before any decision-making process. The process encourages and provides opportunities for the interest groups/stakeholders to share their opinions freely, and for governments to acknowledge them and to find ways to build a bridge between the two entities. Some of the initiatives taken by different interest groups are:

1. **Issue Based Groups**: These are generally urban groups that focus on a particular issue. These groups may interact or indulge in activities carried out by different NGO's or associate with the peoples' movement. Some of these are Toxic Links, LIFE (Legal Initiative for Forest and Environment), Equations (eco-tourism), Samatha (indigenous people), Prayas (electricity regulation), *etc.*
2. **Large Advocacy and Research Organizations:** These are highly structured non-profit organizations that focus on a variety of national and international issues. These are well funded by donor agencies. The agenda of these organizations are not defined by local needs and may work with or against the government for the social benefits of society. Some of these are Centre for Science and Environment, TERI, Greenpeace, *etc.*
3. **People's Movement/ Campaigns:** These are driven by local people due to agitations against local grievances (that starts out as local problems) and thus, are less hierarchical and less structured. Some of the peoples' movements that have gained huge success are the Chipko Movement, Narmada Bachao Andolan, Bhopal Campaign, Right to Information Campaign, *etc.*
4. **Urban NGO's:** These are set up by small groups or individuals, generally driven by volunteers or small staffs for local issues like garbage collection, deforestation, *etc.* The issues are generally taken up in project mode (USEPA, a; Chikkatur, 2007; Hu and Tang, 2014) environmental protection requires public participation. In order to realize the ideals of public participation in environmental protection, we need to enhance the democratization of national administrative agencies, guarantee the civil environmental rights, balance the environmental interests of different groups in light of environmental justice, promote public environmental awareness and cultivate the self-governing spirit in environmental protection. As there are definitely many problems in the public participation of environmental protection, we need to strengthen education in environmental protection, promote public participation in environmental protection and provide support to the civil environmental organizations so as to realize public participation in environmental protection.

The concept of public participation aims at addressing views of different interest groups for a common cause pertaining to environment and sustainability that builds faith and develops sense of integrity for the concept. Also, bringing the voice of community into decision making improves the quality and legitimacy of the process and simultaneously reduces the conflict (Berry *et al.*, 2019) increases the legitimacy of decisions, and helps ensure that policy-makers have valuable local knowledge. Public participation in environmental decision-making includes both formal participation processes (Richardson and Razzaque, 2006). However, there are certain key challenges in conducting meaningful public participation, some of which are discussed below.

- Need for Environment Education. Understanding what Environment and Development mean and converting the understanding to implementable policies (Chikkatur, 2007)
- Involvement of wide cross section of society, particularly urban middle class (USEPA, a).
- Broader public needs for serious engagement in environment debates (USEPA, a).
- Participation may get influenced by those with more power and resources (Mosse, 1994; Agarwal, 1997; Newig, 2007).
- Many people also self-censor because they perceive science as having more authority than they do (Berry *et al.*, 2019) increases the legitimacy of decisions, and helps ensure that policy-makers have valuable local knowledge. Public participation in environmental decision-making includes both formal participation processes (Richardson and Razzaque, 2006).
- Lack of transparency in decision making may lead to "participation fatigue" (Berry *et al.*, 2019) increases the legitimacy of decisions, and helps ensure that policy-makers have valuable local knowledge. Public participation in environmental decision-making includes both formal participation processes (Richardson and Razzaque, 2006).
- Time and Money requirements (Neamtu, Dragos and Capraru, 2014)

In nutshell, integrating and conducting public participation into practice is very challenging due to the aforementioned hurdles. There is an inherent tendency of recreating existing inequalities from existing participation mechanisms. Thus, the need of the hour is to look into the possible solutions for overcoming the above discussed challenges in expanding public participation. Some of the possible solutions discussed by researchers in previous studies are

- Recognizing citizen's action competence in solving environment related issues and promoting education to develop a sense of interrelatedness between man and environment.
- Quick dissemination of useful and reliable information to the public for timely adoption of mitigation measures or policies, particularly, in case of pollution episodes.
- The authorities at local level shall develop policy measures by considering both environmental protection and addressing public views (Hu and Tang, 2014)
- Enabling knowledge sharing, policy making, reflecting the interests of public by appreciating their inputs (Lema-blanco, Garcia-mira and Dumitru, 2014).
- Allowing more transparency by enhancing knowledge of how public opinion can result in shaping up of policies and several decision-making processes (Berry *et al.*, 2019)

Thus, to conclude, public inputs play a pivotal role in developing a positive vision and further collaborating with key stakeholders for achieving sustainable environment. Also, engagement of all the sectors of the society would allow the citizens to act more sustainably.

10.5 Activities and Programs to Promote Environment Education and Public Participation

Rapid urbanization and enhanced pace of developmental activities have overburdened the existing resources. The two major components of the environment; natural (atmosphere, hydrosphere, lithosphere, and biosphere) environment and human-developed environment are facing potential threats due to anthropogenic activities that severely affects its biological and physical counterparts. These activities mostly cause detrimental impacts on soil, water, air quality and quite often leads to adverse climatic changes. Thus, the significance of resource conservation and its components along with the benefits to mankind highlights the importance of environmental education. It is against the backdrop of never-ending demands, human negligence and casual attitudes towards the environment that the concept of Environment Education evolved. The active participation and coordination of the public is an integral component that can't be neglected (Nirmal and Shukla, 2012; Mohamed, 2014). Environmental awareness is imperative and adequate to minimize these problems and achieving environmental sustainability. Public Participation has, therefore, become an indelible feature in environmental decision-making processes (Mohamed, 2014). Thus, in light of the importance of environment protection and the role of different stakeholders in averting the adverse consequences associated with overexploitation of natural resources and environment degradation, several activities and programs have been initiated and conducted by government and/or individuals at micro, macro, national and/or international levels for highlighting the need of environment conservation and regulation. Some of these are discussed in the subsequent sections.

10.5.1 Initiatives Taken by the Indian Government

Environment Information System (ENVIS), 1982: Realizing the importance of disseminating Information related to Environmental Sector, a Scheme has been implemented by the Ministry of Environment, Forest & Climate Change (MoEF&CC) called Environmental Information System (ENVIS). The initiative aims at collecting, collating storing and retrieving related information. In December 1982 (Sixth Plan), ENVIS was initiated for the very first time with prime focus on providing necessary guidance to different stakeholders for better decision making. ENVIS has further collaborated with several agencies/ institutes working in environmental sector. Further, it itself has many nodes known as ENVIS Centers with a Focal Point in the Ministry of Environment & Forests that work in achieving different objectives. Moreover, ENVIS is known as the National Focal Point (NFP) for INFOTERRA which is a global environmental information network of the United Nations Environment Program (UNEP) (MoEF&CC, a, b).

Centre for Environment Education (CEE): CEE, headquartered in Ahmedabad was set up in August 1984. It was set up for promoting education and awareness with the view of development of overall environment and strategies associated therewith. CEE also aims at conducting programs, capacity building for increasing environmental education and communication for sustainable future (CEE).

National Policy on Education, 1986: The policy was framed with the view of improving the quality of education at all levels, focusing on scientific and technical knowledge for promoting active public participation, developing nation's integrity and formation of radical educational system (Chikkatur, 2007).

National Environment Awareness Campaign (NEAC): NEAC was launched by the Ministry of Environment, Forest & Climate Change (MoEF&CC) at a national level in order to reconnect with nature. Since 1986, it is conducted every year to promote the importance of environmental education and create awareness among the target groups nationwide for active participation and involvement (Bureau)

Policy Statement for the Abatement of Pollution, 1992: It lays emphasis on the prevention and control of pollution and adopting technically feasible measures (such as solid waste management by waste minimization, 3R principle, water and air quality improvement *etc.*), for achieving sustainable environment (Inernational Centre for Environment Audit and Sustainable Development).

Earth Summit, 1992: It comprises of 27 principles addressing all environmental issues pertaining to sustainability, poverty, emission reductions and active involvement of several target groups for conserving environment for both present and future generation (Neamtu, Dragos and Capraru, 2014).

Regional Centre for Expertise (RCE) on Education for Sustainable Development: It is a centre comprising of different sectors (formal/ informal) with the aim of imparting knowledge on various aspects in order to achieve sustainable future. Some of the work areas of RCE are

- Access to quality education
- Conducting programs targeting environment and conserving nature for sustainable development
- Involving masses for promoting environmental education
- Providing technical know-how (Global, Network and Polic).

National Green Corps: It is a program that provides provision of educating children about environment related issues through eco-clubs initiated in the year 2001-02. NGC Programme builds cadres of the children and youth involved actively in conserving environment and nature. Also, it encourages children to come up with their own views, opinions and creativities for protecting environment and betterment of the society as a whole (CPREEC)

National Museum of Natural History: The museum is an informal way of teaching and inculcating a sense of interest in nature's creations. Late Smt. Indira Gandhi developed the concept on 15 August 1972. The Museum excels in attracting the young minds through its antique pieces, galleries, theme based concepts *etc.* It also sensitize people towards the environment and builds awareness among the masses (MoEF&CC, c).

Swacch Bharat Abhiyan: It is a massive movement launched by Narendra Modi on 2 October 2014 with a goal of achieving Clean India by 2019 by educating people in both urban and rural areas. The program focusses on open defecation free country within five years, proper sanitation, waste management and bringing attitudinal changes within the people for promoting healthy practices in both urban and rural areas for a sustainable future (Vikaspedia).

National Action Plan on Climate Change: It is a policy document prepared by Prime Minister's Council on Climate Change in 2008. It includes eight national missions addressing solar energy, water and energy conservation, sustainable habitat, ecosystem, forest management, afforestation and initiatives towards Green India (NeoStencil).

10.5.2 CEPA Action Plan

CEPA is **C**ommunication, **E**ducation and **P**ublic **A**wareness that mainly revolves around processes and actions for cross cutting environmental issues with the prime focus on biodiversity. Its main aim is to conserve the biological resources by involving different target groups or stakeholders and educating them for the same. CEPA therefore, act as a tool to develop capacity (networks, partnerships, and knowledge management) to support and promote environmental sustainability. It even provides a means to gain the cooperation of different interest groups and encourage participatory dialogue. It is basically on the concept of adaptive management that helps to learn from experiences. The words associated with CEPA acronym were expanded by the IUCN Commission on Education and Communication to expose the range of tools and processes involved in bringing about change in people and society (Fig. 3).

C	**E**	**P**	**A**
Communications Connection Changes in Behavior Capacity Building	Education Empowerment	Public Awareness Public Participation	Action Action Research

Fig. 3: Basic Concept of CEPA (Hesselink *et al.*, 2007)

C

Communication: It establishes a sense of understanding amongst different stakeholders through information exchange for effective planning for protecting environment.

Capacity Building: Develops networks, partnerships to enhance the ability of interest groups to collaborate and work for the common cause of environment.

E

Education: Imparts knowledge, motivation and skills about the issues and concerns pertaining to environment and simultaneously clarifies values, develops attitudes of concern towards environment.

Empowerment: Involvement for effective decision making.

P

Public Awareness: Sensitize people about the importance of environment, related issues and the need for protection for sustainable future.

Public Participation: Allows masses to participate and come up with their view points for working collectively for the betterment of the environment.

A

Action/ Action Research: Aims at bringing changes based on experiences to conserve environment for sustainable development.

CEPA National Focal Points

CEPA also comprises of CEPA National Focal Points that are recognized experts (one official from government and one from an NGO). They work in collaboration for conducting related programs. Some of their key roles are discussed below:

- Providing conducive environment for different stakeholders to work
- Assistance in nationwide implementation of programs under CEPA
- Goal setting for conserving biological resources
- Network building, information exchange, education, communication for effective implementation (RAMSAR).

Policy Phases and Roles of CEPA

CEPA plays several roles in different phases of policy processes for educational communication and nationwide awareness and participation. The policies aims at modifying already followed practices and adopting new measures for conservation and judicious use of biological resources. The same has been described in Fig. 4.

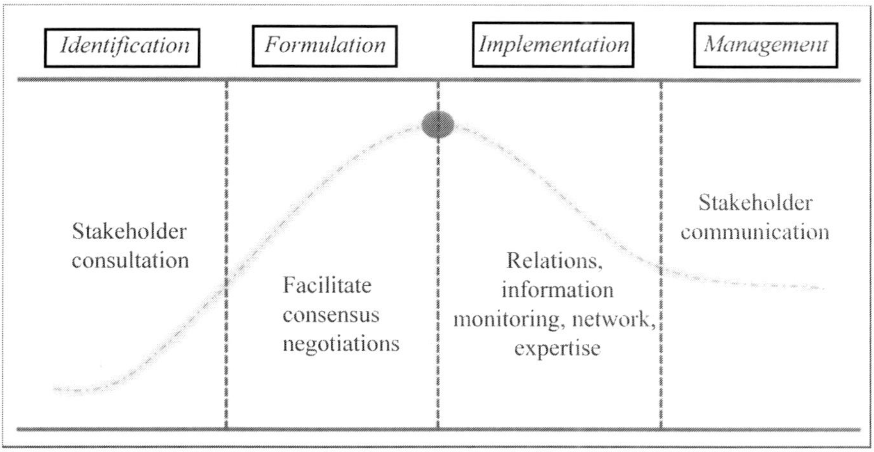

Fig. 4: Interactive Communication in all policy phases (Hesselink *et al.*, 2007)

Identification Phase: The role and responsibilities of different organizations/ groups/ active members in achieving the agenda are addressed during Identification Phase. The different methods used for the same are meetings on regular intervals, wide-networking, information exchange, individual views and opinions *etc*.

Policy Formulation: It involves mass awareness of the environment related problems to seek support from different agencies/ means for consulting and adopting appropriate measures/ policies for protecting natural resources.

Policy Implementation: The implementation phase involves imparting education on different policy measures to educate the target groups regarding the benefits for capacity building.

Management and control: In this phase, the adopted management strategies are briefed to obtain feedback from different member groups and stakeholders (RAMSAR)

Thus, CEPA is a collaboration and cooperation of stakeholders and beneficiaries that act as drivers of the society in a planned and systematic approach to reflect their varied interests, knowledge, and network. In view of this, the above discuss strategy framework has been prepared for developers and practitioners for facilitating environment conservation with the support of multi-stakeholders.

10.6 International Involvement and Collaboration: A new Vision

The extremities of environmental conservation involves the fact that the impact of anthropogenic activities must be measured and assessed for its adverse consequences, policies and strategies. Most of the international agreements, conventions, and collaboration for the environment and its protection aim towards these objectives for conserving nature in order to attain a sustainable future. Recognizing the importance and need to safeguard natural resources, there is a need for the global participation of the nation for effective environment regulation.

Since historical times, globally, India has been actively involved in conservation practices/ treaties/ conventions/ protocols that supported improved environmental outcomes. India even signed CITES, Convention on International Trade in Engendered Species of Flora and Fauna for regulation, prevention, and control of trade in endangered species. India adopted the Montreal Protocol aimed at reducing ODS, i.e., Ozone Depleting Substances in September. In 1989, India ratified Basel Convention on Transboundary Movement of Hazardous Waste. In 1992, India became a member of UNFCCC and ratified in 1993. The convention aimed at regulating greenhouse gas emissions, thereby bringing emissions to a level that could offset the adverse climatic changes and global warming. According to UNFCCC, as a developing nation, India was not bound to the commitments, but with Paris Agreement at COP21, India committed to participate in multilateral negotiations under UNFCCC where National Action Plan on Climate Change was adopted. India is also a signatory to Agenda 21 that aims at Sustainable Development. India is even party to Cartagena Protocol on Biosafety (2000) (*Geography and You*, 2018).

EPA's engagement with India has contributed significantly in addressing critical environmental issues. For the years 2005-2011, EPA collaborated with MoEF&CC for capacity building In India to look after environmental Compliance and Enforcement. In an initiative to combat climate change, Global Methane Initiative (GMI) was joined by India that aims at reducing methane emissions and advancing abatement, recovery, and use of methane as a valuable energy source. In March 2010, New Delhi hosted GMI's highly successful Partnership Expo for promoting technologies for recovering methane gas. In April 2014, India with support of EPA published a handbook on "Enforcing Hazardous Waste Rules in India: Strategies and Techniques for Achieving Increased Compliance". In 2015, a pledge was taken by US and India to cooperate on clean energy along with climate change. India even participated in the International E-Waste Management Network that enabled

Indian officials to participate in the US and Taiwan. It provided them with a platform to share experiences and knowledge (USEPA, a).

Recently, in March 2019, UNEP and Japan have announced to develop counter measures against marine plastic litter in India (UNEP). Also, in context with environmental sustainability, India-Africa Partnership has grown exponentially, which can also be witnessed through the International Solar Alliance initiative adopted in recent past in 2015 at COP21. US$2 billion credit to Africa for five years was pledged by India for developing 10,000 MW solar power systems. Recently, on 15 March 2019, Centre for New Economic Diplomacy hosted roundtables, where-in efforts were made for fostering relationships between India and African nations in implementation of Sustainable Development Goals.

Thus, recognizing the significance of international agreements/ conventions/ treaties/ collaborations/ cooperation/ coordination, it can be easily deduced that India's partnership in recent past with different nations worldwide has led to a tremendous progress in protecting environment and conserving biological resources for a sustainable future.

References

Agarwal, B. (1997) 'Editorial: Re-sounding the alert—gender, resources and community action', World Development, 25(9), pp. 1373–1380. doi: 10.1016/s0305-750x(97)00062-4.

Alexandar, R. (2014) 'Alexandar-Poyyamoli-JSE-Vol-7-Dec2014', 7(December).

Asian Age (2017) 'UGC pushes for compulsory environmental studies'. Available at: https://www.asianage.com/india/all-india/230617/ugc-pushes-for-compulsory-environmental-studies.html.

Berry, L. H. et al. (2019) 'Making space: how public participation shapes environmental decision-making', pp. 1–8. Available at: https://www.sei.org/publications/how-public-participation-shapes-environmental-decision-making/?utm_content=buffer5e5b7&utm_medium=social&utm_source=facebook.com&utm_campaign=buffer.facebook.

Bhat, S. A. et al. (2017) 'Environmental Education in India: An Approach to Sustainable Development', FIIB Business Review, 6(1), pp. 14–21. doi: 10.1177/2455265820170102.

Brian A. Day, M. C. M. (2000) Environmental Education & Communication for a Sustainable World.

Bureau, P. I. (no date) Challenges to Reconnect with Nature. Available at: https://pibindia.wordpress.com/2017/06/05/challenges-to-reconnect-with-nature/#more-12028.

CEE (no date) Centre for Environment Education, About Us. Available at: https://www.ceeindia.org/about-us.

Chang, T. (2016) 'Infusing Environmental Education into National Curriculum Framework in Taiwan'.

Chikkatur, A. (2007) 'Environmental Governance in India Issues and Challenges', History. CPREEC (no date) National Green Corps (NGC). Available at: http://www.cpreec.org/special-national.htm.

Development, I. C. for E. A. and S. (no date) Policy Statement for the Abatement of Pollution, 1992. Available at: http://iced.cag.gov.in/?page_id=1034.

Festus, M. O. and Ogoegbunam, O. B. (2012) 'Imperatives of Environmental Education and Awareness', Academic Research International, 3(2), pp. 253–259.

Foundation, I. Y. F. (2008) Youth and the Environment.

Geography and You (2018) 'International Agreements On The Environment And India', 21 April. Available at: https://www.geographyandyou.com/economy/international-agreements-environment-india/.

Global, T. H. E., Network, R. C. E. and Policy, E. S. D. (no date) 'The Global RCE Network and ESD Policy'.

Gopal, G. V. and Anand, V. V. (2008) 'Environmental Education in school Curriculum an overall perspective', Convenor Environmental Information System (ENVIS), 22, pp. 44–58.

Hesselink, F. et al. (2007) 'Communication, education and public awareness (CEPA): a toolkit for national focal points and NBSAP coordinators', p. 310pp. Available at: http://www.cbd.int/cepa/toolkit/2008/doc/CBD-Toolkit-Complete.pdf.

Hu, X. H. and Tang, L. (2014) 'Public participation in environmental protection: Ideals, challenges and solutions', Advanced Materials Research, 962–965, pp. 1680–1684. doi: 10.4028/www.scientific.net/AMR.962-965.1680.

India Today (2017) 'Supreme Court directs universities to include Environment Studies in curricula', 7 March. Available at: https://www.indiatoday.in/education-today/news/story/government-schools-964295-2017-03-07.

ISA (2020) First World Solar Technology Summit.

ISWA (no date) 'Environmental Communication for municipal solid waste management inSao Paulo, Brazil'.

Khademi-Vidra, A. (2017) 'Curriculum Planning of the Environmental Education', Journal of Central European Green Innovation, 5(1), pp. 65–80.

Lema-blanco, I., Garcia-mira, R. and Dumitru, A. (2014) 'Environmental education , public participation and Bottom-Up Processes .', (June).

Malley, S. O. (2020) '(Re) Connecting Children with Nature ? A Sociological Study of Environmental Education in Ireland', 340406970(November 2014). doi: 10.13140/RG.2.2.26435.55844.

MoEF&CC (2017) Graded Action Plan to reduce Urban Air Pollution. Available at: https://pib.gov.in/newsite/PrintRelease.aspx?relid=159542.

MoEF&CC (no date a) Environmental Information System. Available at: http://envis.nic.in/ENVIS_html/about.html.

MoEF&CC (no date b) Environmental Information System. Available at: http://envis.nic.in/content/AboutUs.aspx.

MoEF&CC (no date c) National Museum of Natural History. Available at: http://www.nmnh.nic.in/aboutus.htm.

Mohamed, E. (2014) 'Environmental Education and Public Awareness', Journal of Educational and Social Research, (March). doi: 10.5901/jesr.2014.v4n3p333.

Mosse, D. (1994) 'Authority, Gender and Knowledge: Theoretical Reflections on the Practice of Participatory Rural Appraisal', Development and Change, 25(3), pp. 497–526. doi: 10.1111/j.1467-7660.1994.tb00524.x.

Neamtu, B., Dragos, D. C. and Capraru, L. (2014) 'Public Participation in Environmental Decision Making in Romania', Central European Public Administration Review, 12(2–3), pp. 63–80. doi: 10.17573/ipar.2014.2-3.a04.

NeoStencil (no date) National Action Plan on Climate Change. Available at: https://neostencil.com/national-action-plan-on-climate-change.

Newig, J. (2007) 'Does public participation in environmental decisions lead to improved environmental quality', CCP (Communication, Cooperation, Participation. Research and Practice for a sustainable Future), 1(October 2001), pp. 51–71. Available at: http://www.usf.uni-osnabrueck.de/~jnewig/03_Forschung_Newig_final.pdf.

Nirmal, B. C. and Shukla, P. C. (2012) 'Public Participation in Environmental Decision Making : International Standards and Indian State Practice 1', The Banaras Law Journal, 41(2), pp. 9–46.

Pradhan, H. (2019) 'UGC pushes for compulsory environmental studies', The Times of India, 17 May. Available at: https://timesofindia.indiatimes.com/city/bhubaneswar/ugc-for-compulsory-environment-studies/articleshow/69363905.cms.

Praveen, H. and Nasreen, N. (no date) 'Status of Environmental Education at Secondary School Level', 2(1).

Ramchandra Acharya (2013) Role of Youth for A Cleaner and Greener Environment. Available at: https://ypard.net/2013-june-25/role-youth-cleaner-and-greener-environment.

RAMSAR (no date) CEPA Focal Points in your Country.

Richardson, B.J. and Razzaque J. (2006) 'Public participation in environmental decision-making', Environmental law for sustainability, 6, pp. 165-94

Sayal, R. et al. (2016) 'Fostering Systems Thinking for Youth Leading Environmental Change: A Multinational Exploration', Ecopsychology, 8(3), pp. 188–201. doi: 10.1089/eco.2016.0023.

Sharma, P. K. (no date) Case Study Spreading Climate Change Literacy on Wheels: Science Express – Climate Action Special.

Sonowal, C. J. (2009) 'Environmental Education in Schools: The Indian Scenario', Journal of Human Ecology, 28(1), pp. 15–36. doi: 10.1080/09709274.2009.11906215.

UNESCO (no date) Education for Sustainable Development. Available at: https://en.unesco.org/partnerships/partnering/education-sustainable-development.

USEPA (no date a) International Cooperation. Available at: https://www.epa.gov/international-cooperation/public-participation-guide.

USEPA (no date b) What is Environmental Education? Available at: https://www.epa.gov/education/what-environmental-education.

Vikaspedia (no date) Swachh Bharat Mission. Available at: http://vikaspedia.in/health/sanitation-and-hygiene/swachh-bharat-mission.

Wals, A. (no date) Case Study NatureWise: Dutch Children ' s Learning In , About , For , and By Nature.